In response to seeing that Mr. Stewart had published "P(h)osers" as his first "Adventures in AI Usage" featuring what free AI will do with female subjects, or wants to do, or is programmed to do, I decided to create this look into how free AI responds to prompts for images of male Pumping Posers, under the title "P(h)umping P(h)osers". Strange that there were still, (what should I consider them?) 'child safety restraints' on use of language in prompts and limitations on what words were allowed to be pictured, even with male subjects. I will include more details in our upcoming book of "Conclusions -Adventures in AI Usage"

In Text to image

well-built, beautiful man in well-filled speedo pumping gas into a beautiful car at gas station

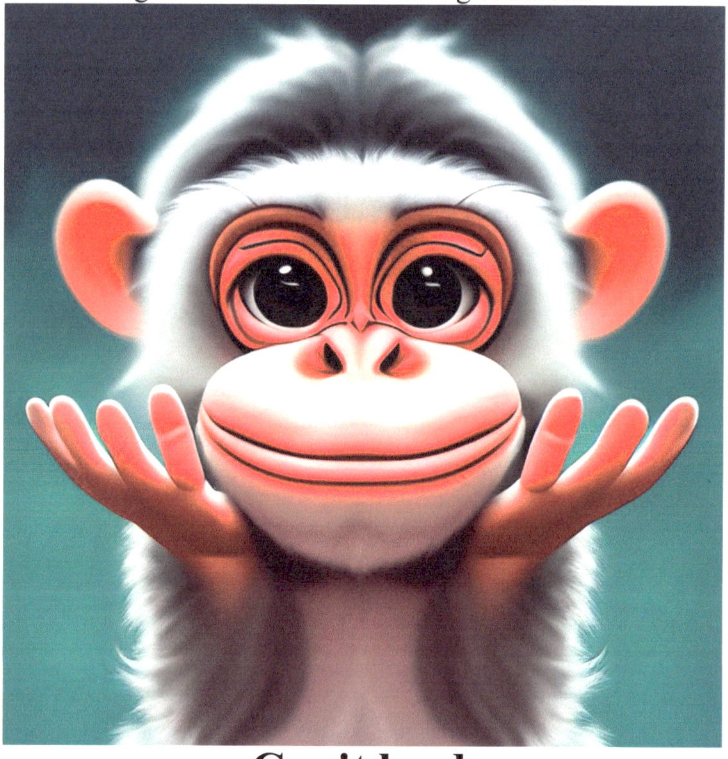

Can't load

We can't display the generated image.
You can edit and try again.

well-built, beautiful man in speedo pumping gas into a beautiful car at gas station

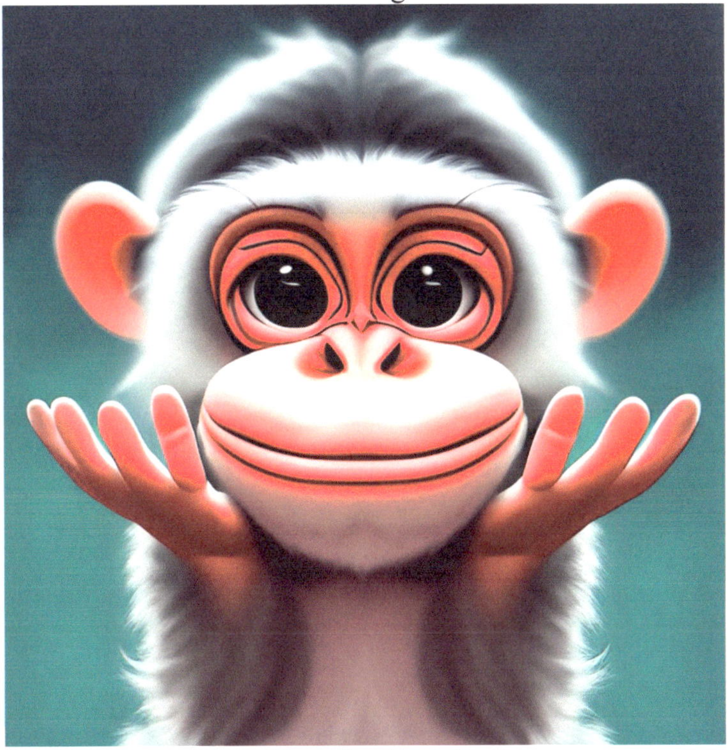

Can't load

We can't display the generated image.
You can edit and try again.

\ man in speedo pumping gas into a beautiful car at gas station

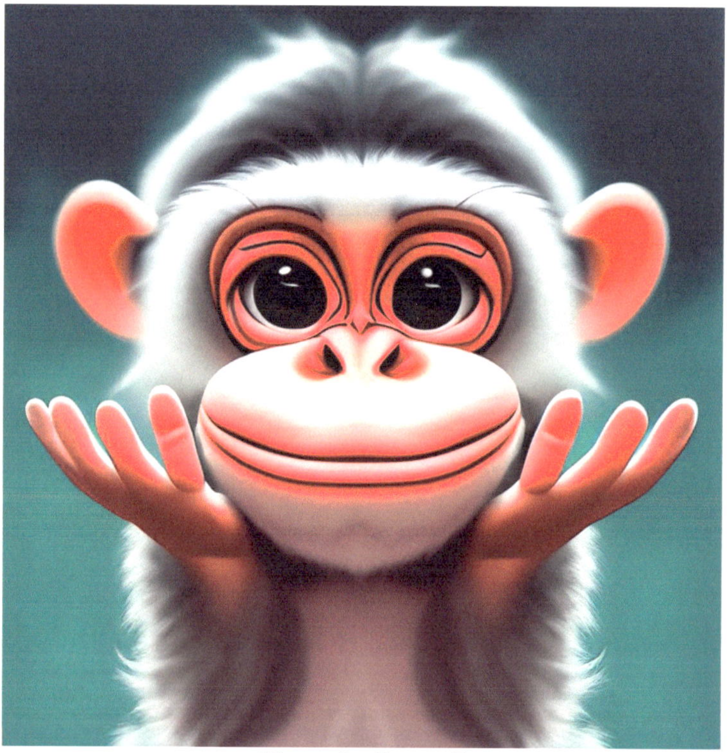

Can't load

We can't display the generated image.
You can edit and try again.

well-built, beautiful man in well-filled speedo pumping gas into a beautiful car at gas station

well-built, beautiful man in well-filled speedo pumping gas into a beautiful car at gas station

well-built, beautiful man in well-filled speedo pumping gas into a beautiful car at gas station

well-built, beautiful man in well-filled speedo pumping gas into a beautiful car at gas station

well-built, beautiful man in well-filled speedo pumping gas into a beautiful car at gas station

well-built, beautiful man in well-filled speedo pumping gas into a beautiful car at gas station

well-built, beautiful man in well-filled speedo pumping gas into a beautiful car at gas station

well-built, beautiful man in well-filled speedo pumping gas into a beautiful car at gas station

well-built, beautiful man in well-filled speedo pumping gas into a beautiful car at gas station

well-built, beautiful man in well-filled speedo pumping gas into a beautiful car at gas station

well-built, beautiful man in well-filled speedo pumping gas into a beautiful car at gas station

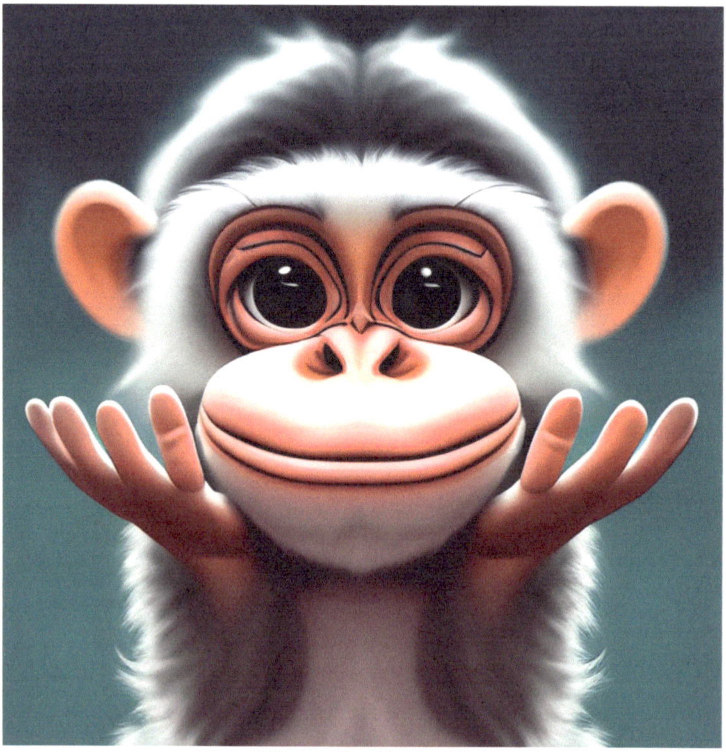

Can't load

We can't display the generated image.
You can edit and try again.

shot from above, well-built, beautiful man in well-filled speedo pumping gas into a beautiful car at gas station

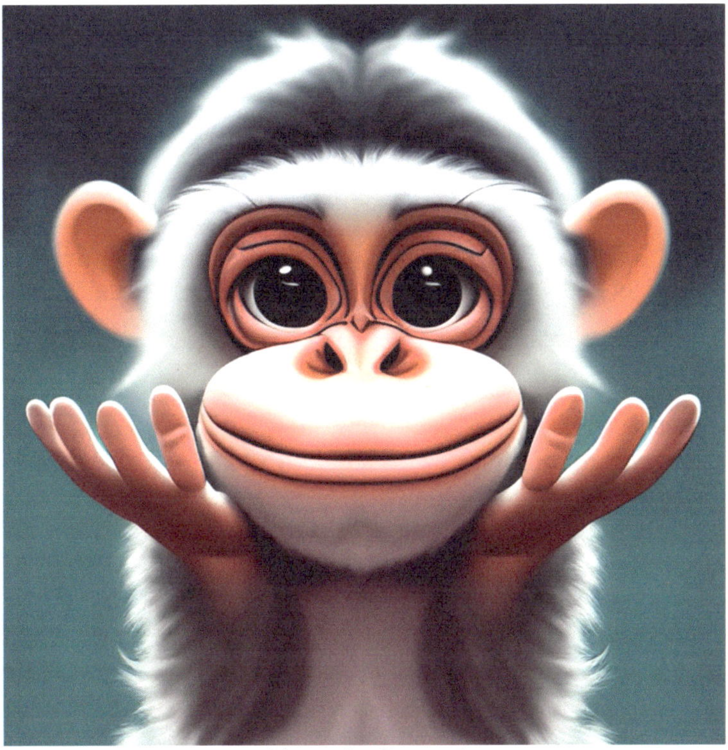

Can't load

We can't display the generated image.
You can edit and try again.

shot from above, well-built, beautiful man in well-filled speedo pumping gas into a beautiful car at gas station

shot from above, well-built, beautiful man in well-filled speedo pumping gas into a beautiful car at gas station

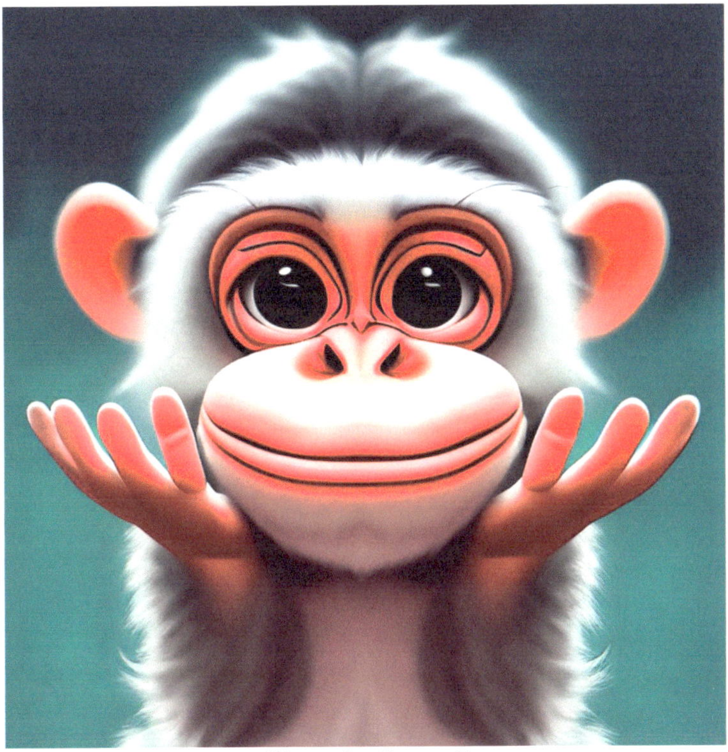

Can't load

We can't display the generated image.
You can edit and try again.

shot from above, well-built, beautiful man in well-filled speedo pumping gas into a beautiful car at gas station

shot from above, well-built, beautiful man in well-filled speedo pumping gas into a beautiful car at gas station

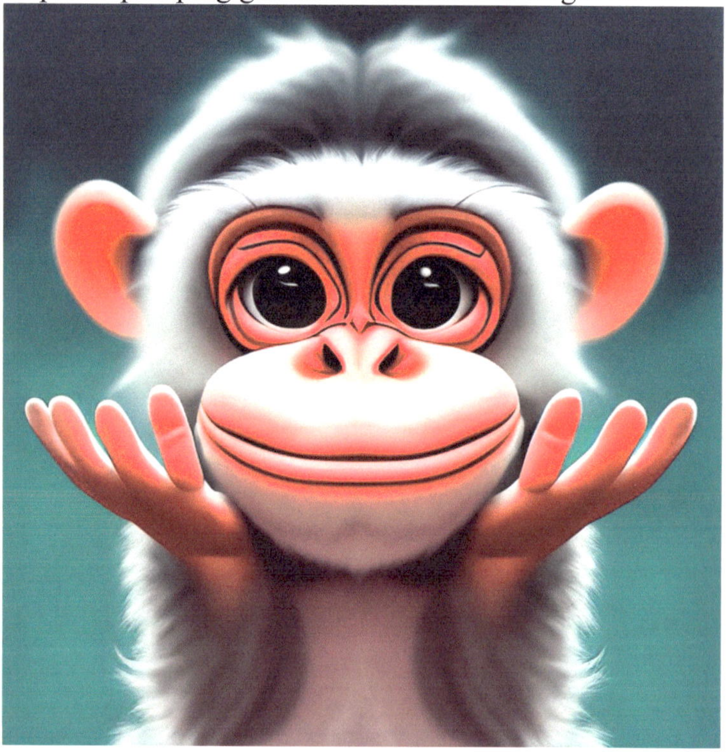

Can't load

We can't display the generated image.
You can edit and try again.

shot from above, well-built, beautiful man in well-filled speedo pumping gas into a beautiful car at gas station

shot from above, well-built, beautiful man in well-filled speedo pumping gas into a beautiful car at gas station

shot from below, well-built, beautiful man in well-filled speedo pumping gas into a beautiful car at gas station

shot from below, well-built, beautiful man in well-filled speedo pumping gas into a beautiful car at gas station

shot from below, well-built, beautiful man in well-filled speedo pumping gas into a beautiful car at gas station

shot from below, well-built, beautiful man in well-filled speedo pumping gas into a beautiful car at gas station

shot from below, well-built, beautiful man in well-filled speedo pumping gas into a beautiful car at gas station

shot from below, well-built, beautiful man in well-filled speedo pumping gas into a beautiful car at gas station

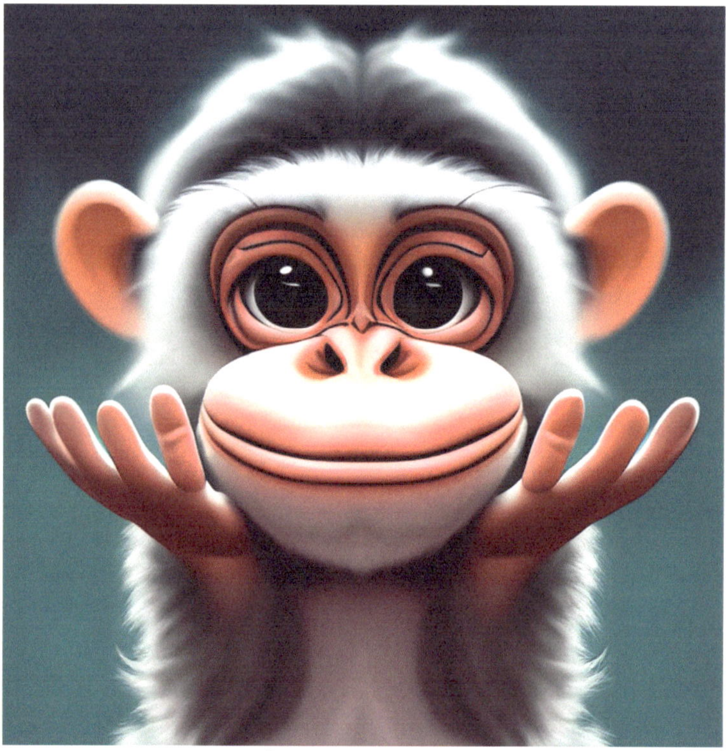

Can't load

We can't display the generated image.
You can edit and try again.

shot from below, well-built, beautiful man in well-filled speedo pumping gas into a beautiful car at gas station

shot from below, well-built, beautiful man in well-filled speedo pumping gas into a beautiful car at gas station

shot from below, well-built, beautiful man in well-filled speedo pumping gas into a beautiful car at gas station

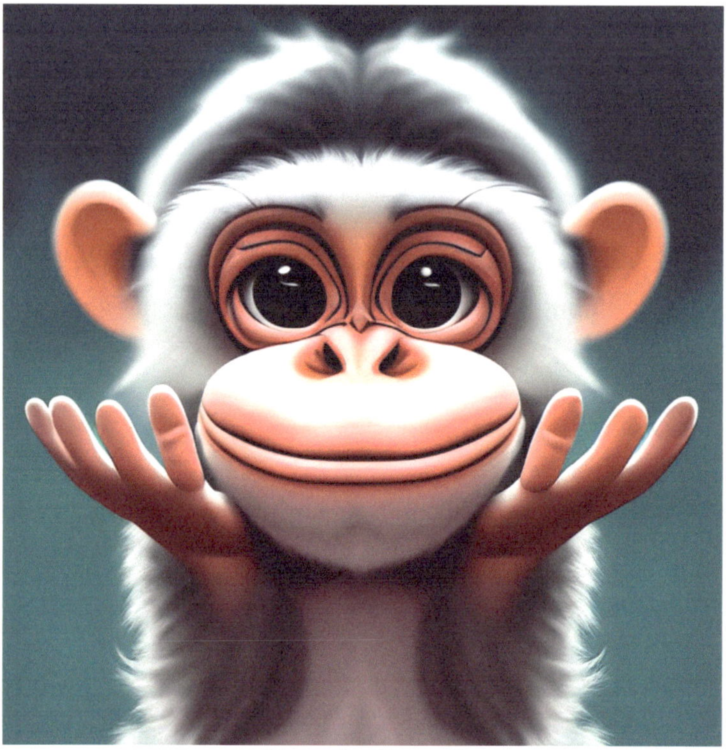

Can't load

We can't display the generated image.
You can edit and try again.

shot from below, well-built, beautiful man in well-filled speedo pumping gas into a beautiful car at gas station

shot from below, well-built, beautiful man in well-filled speedo pumping gas into a beautiful car at gas station

shot from below, well-built, beautiful man in well-filled speedo pumping gas into a beautiful car at gas station

shot from below, well-built, beautiful man in well-filled speedo pumping gas into a beautiful car at gas station

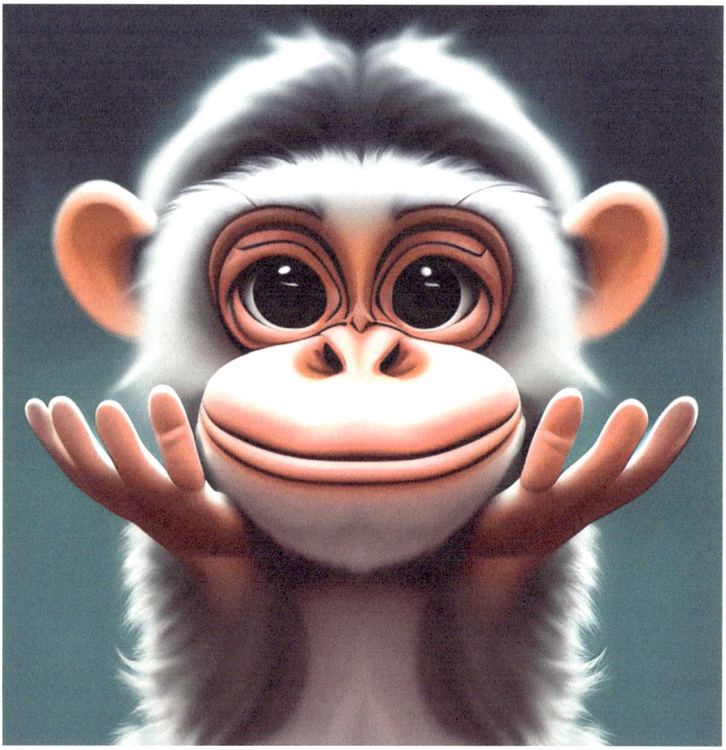

Can't load

We can't display the generated image.
You can edit and try again.

shot from below, well-built, beautiful man in well-filled speedo pumping gas into a beautiful car at gas station

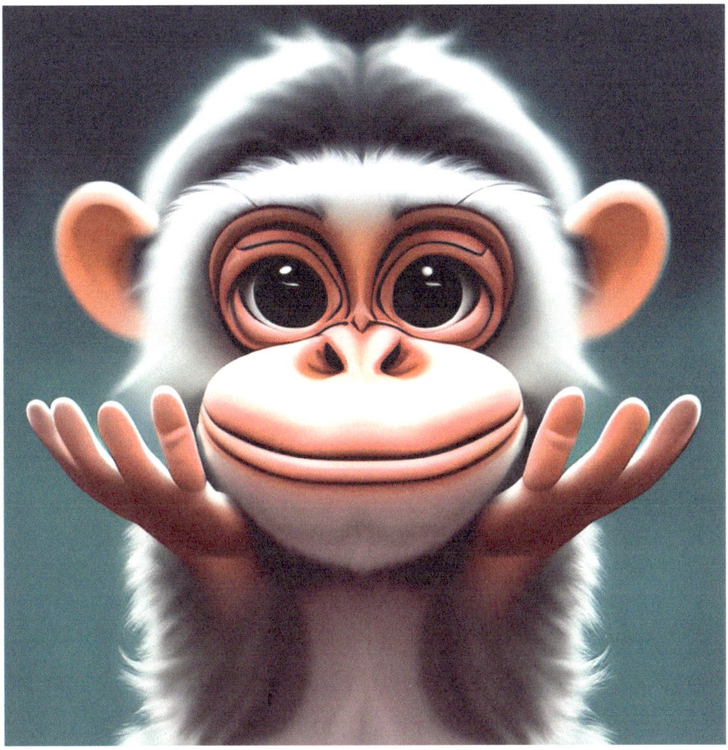

Can't load

We can't display the generated image.
You can edit and try again.

shot from below, well-built, beautiful man in well-filled speedo pumping gas into a beautiful car at gas station

shot from below, well-built, beautiful man in well-filled
speedo pumping gas into a beautiful car at gas station

shot from below, well-built, beautiful man in well-filled
speedo pumping gas into a beautiful car at gas station

shot from below, well-built, beautiful man in well-filled
speedo pumping gas into a beautiful car at gas station

shot from below, well-built, beautiful man in well-filled
lulu lemon pumping gas into a beautiful car at gas station

shot from below, well-built, beautiful man in well-filled
lulu lemon pumping gas into a beautiful car at gas station

shot from below, well-built, beautiful man in well-filled lulu lemon pumping gas into a beautiful car at gas station

shot from below, well-built, beautiful man in well-filled lulu lemon pumping gas into a beautiful car at gas station

shot from below, well-built, beautiful man in well-filled lulu lemon pumping gas into a beautiful car at gas station

shot from below, well-built, beautiful man in well-filled lulu lemon pumping gas into a beautiful car at gas station

shot from below, well-built, beautiful man in well-filled lulu lemon pumping gas into a beautiful car at gas station

shot from below, well-built, beautiful man in well-filled lulu lemon pumping gas into a beautiful car at gas station

shot from below, well-built, beautiful man in well-filled lulu lemon pumping gas into a beautiful car at gas station

shot from below, well-built, beautiful man in well-filled lulu lemon pumping gas into a beautiful car at gas station

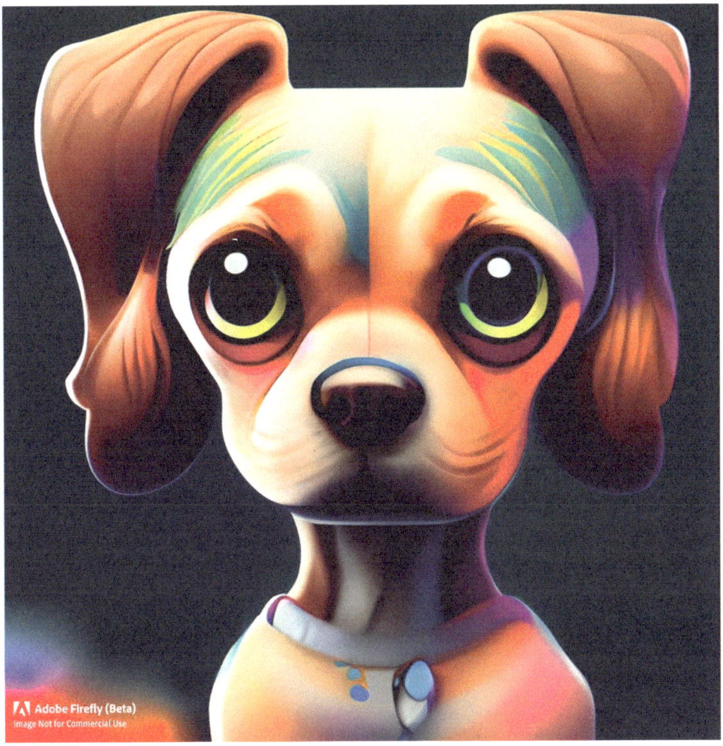

Something went wrong

Please try again later.

shot from below, well-built, beautiful man in well-filled lulu lemon pumping gas into a beautiful car at gas station

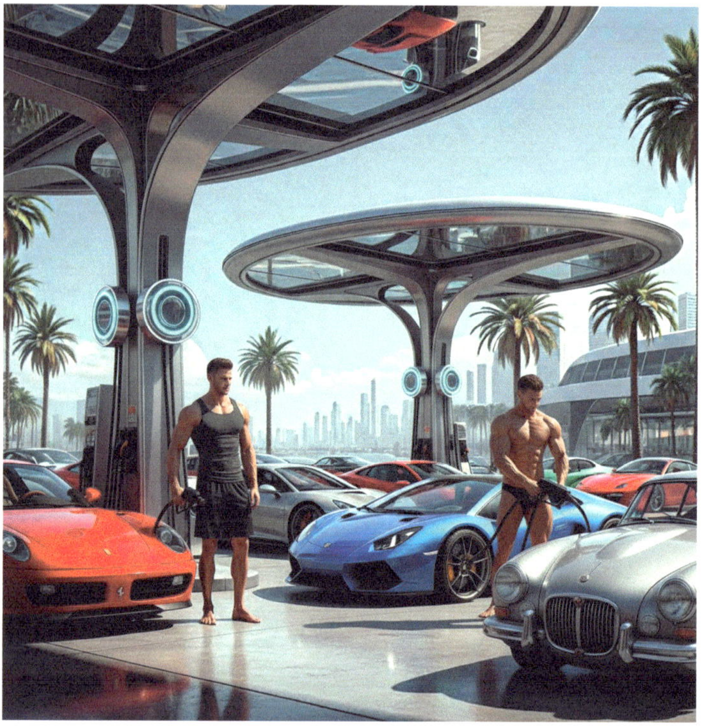

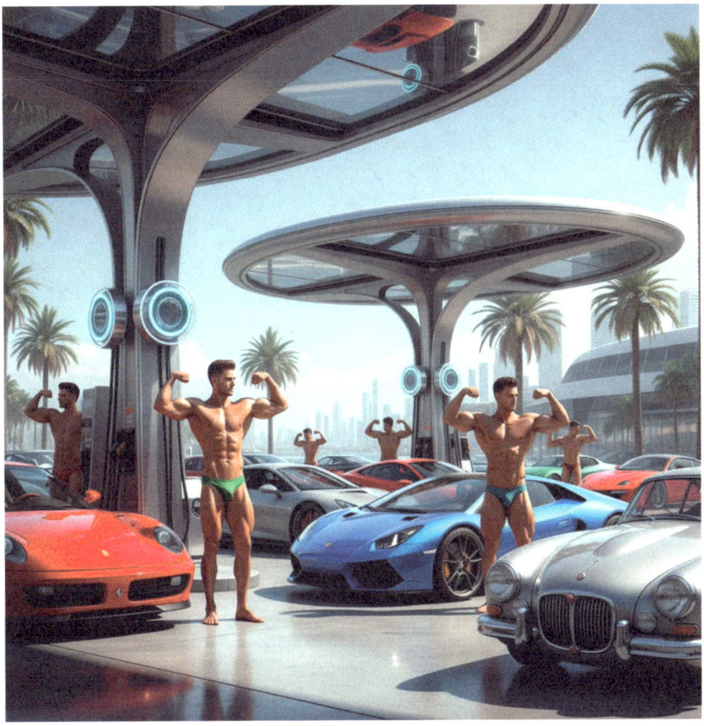

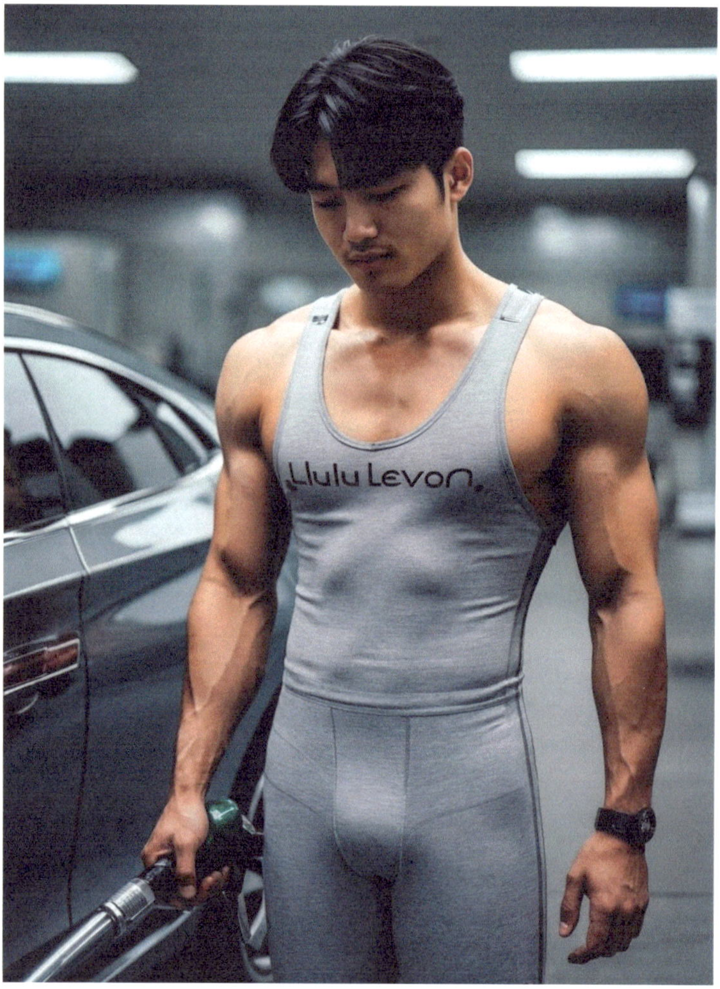
Llulu Levon

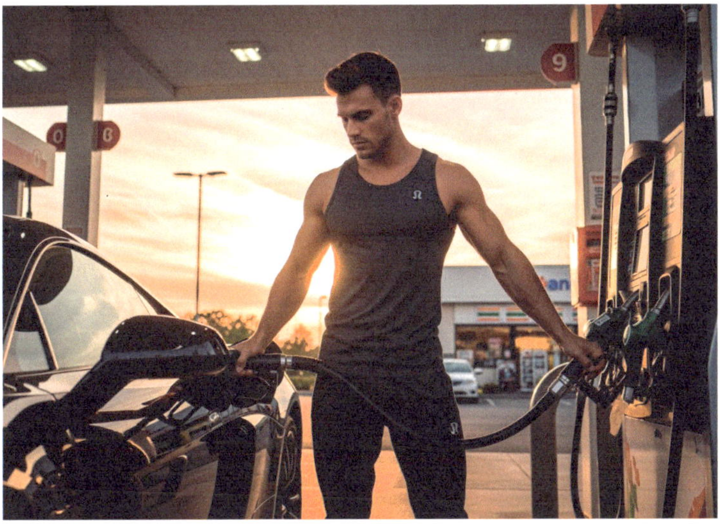
9

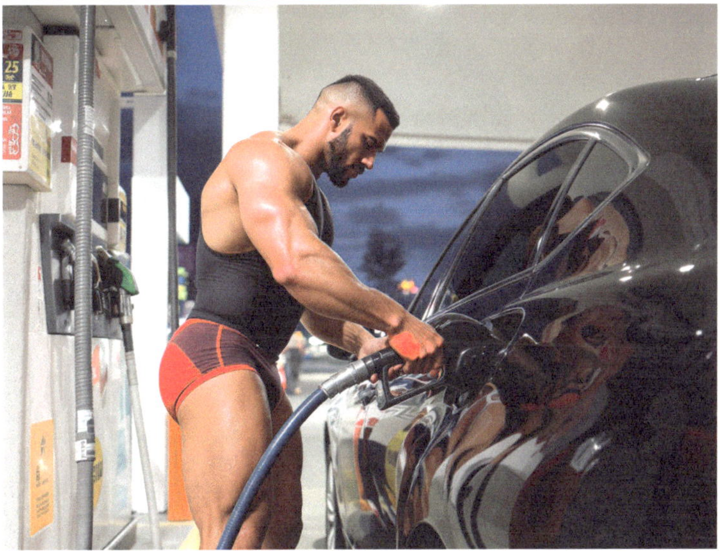
25

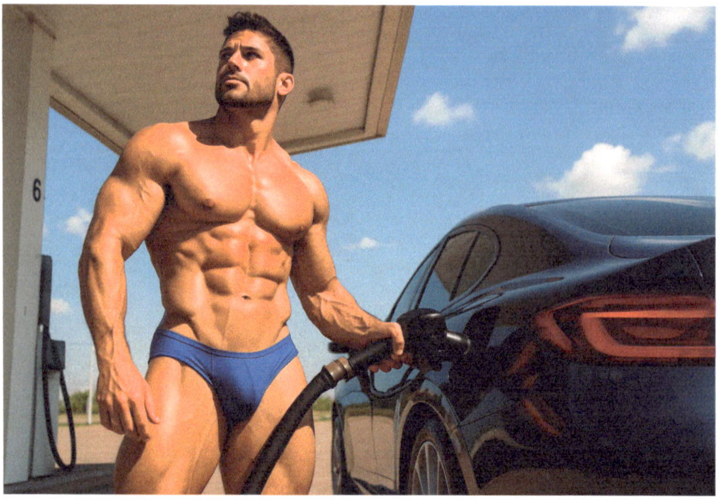
6

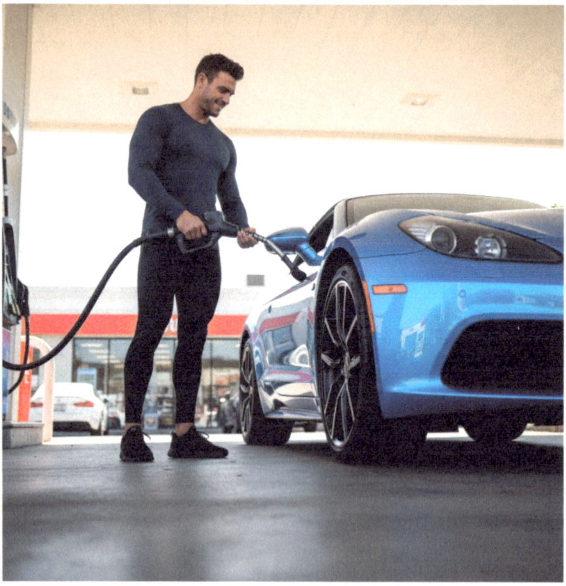

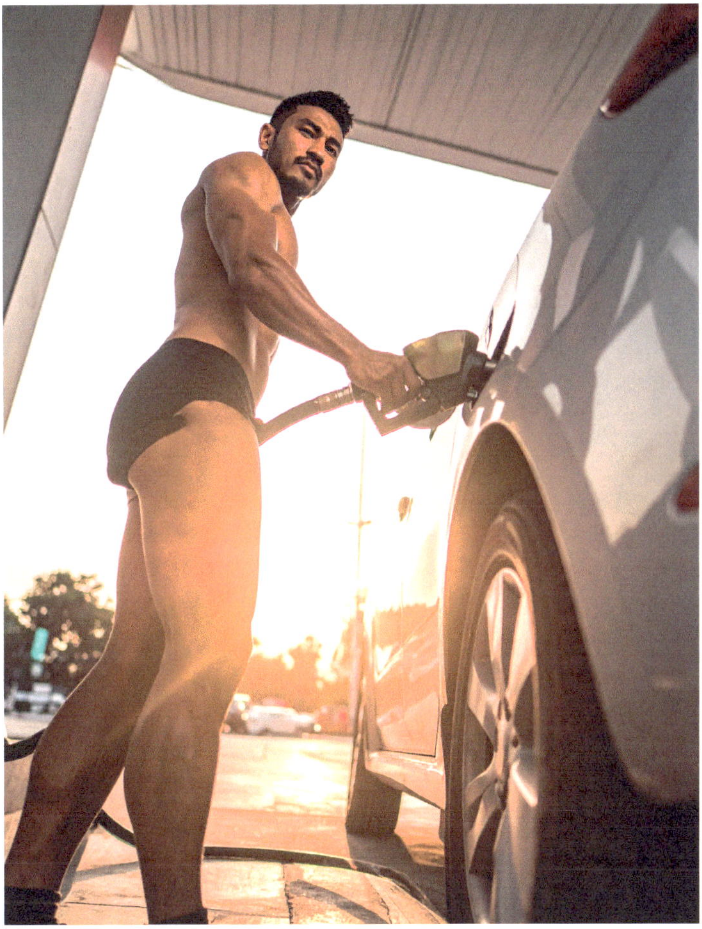

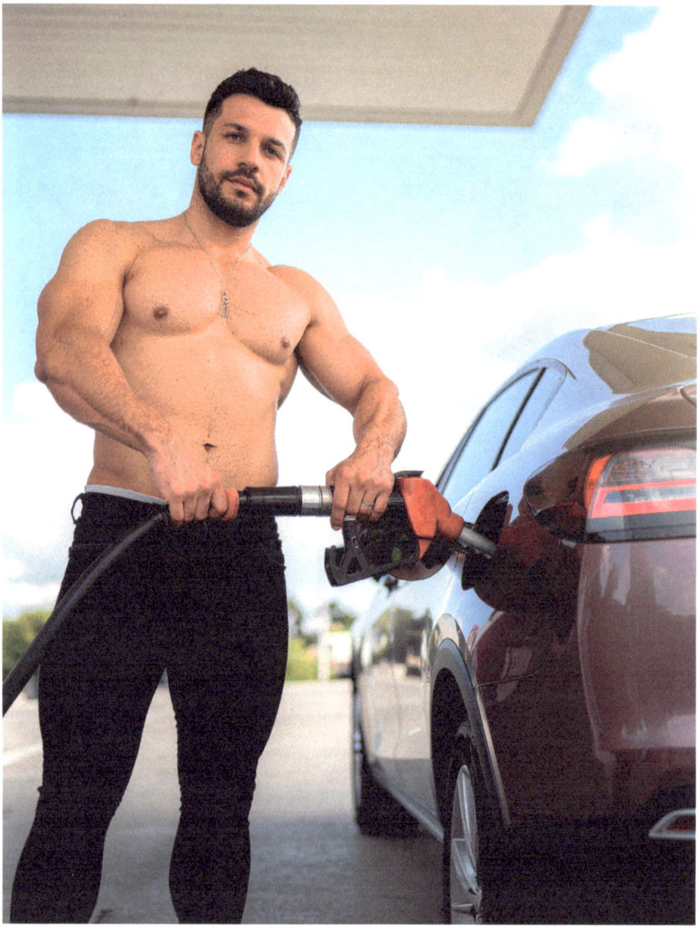

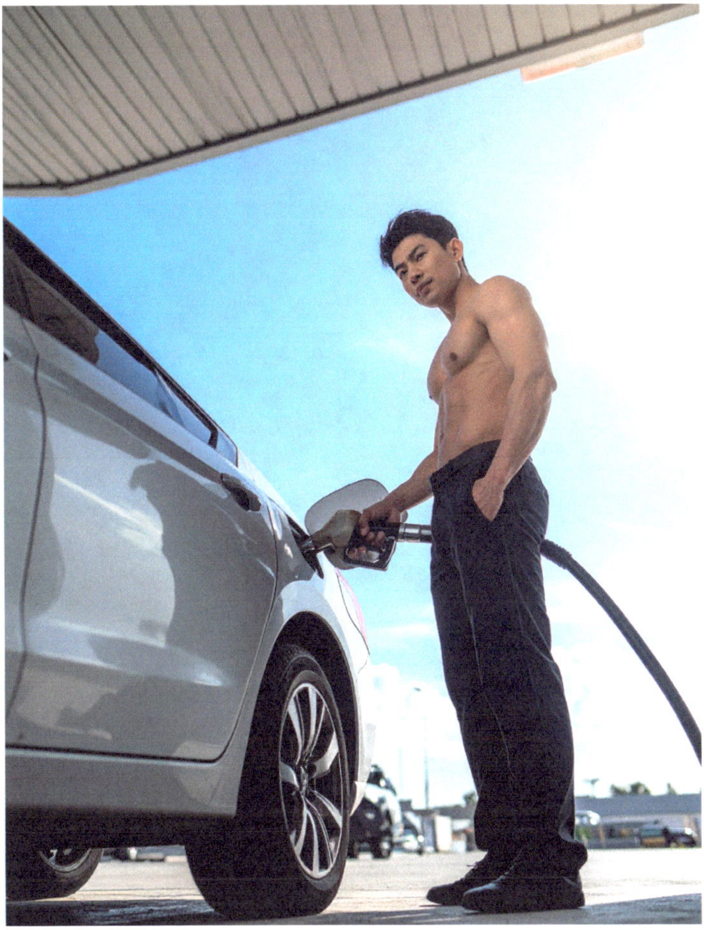

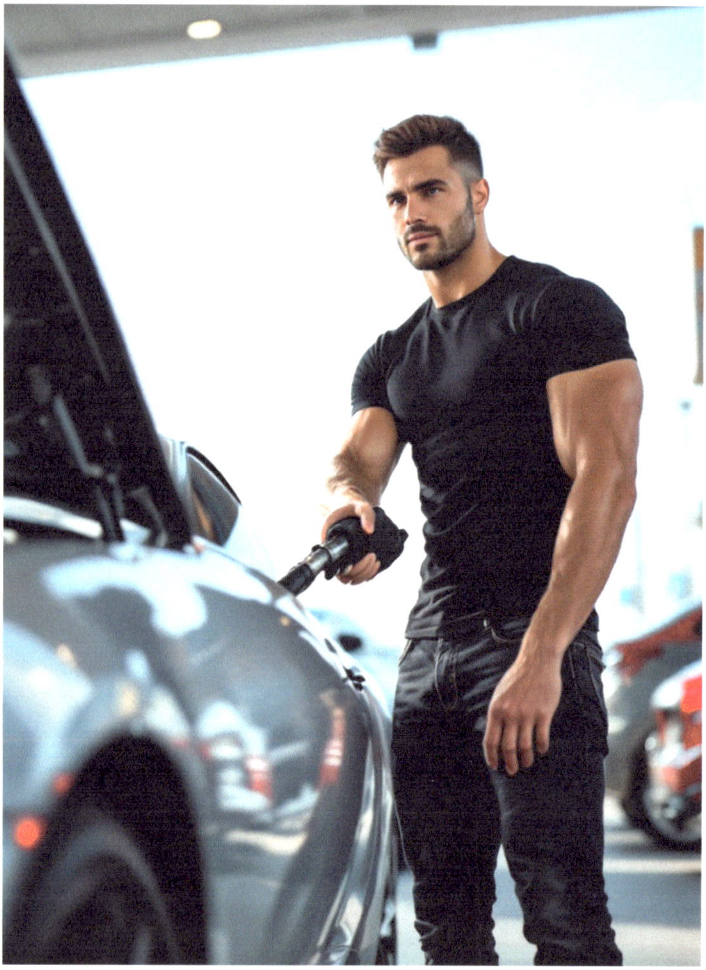

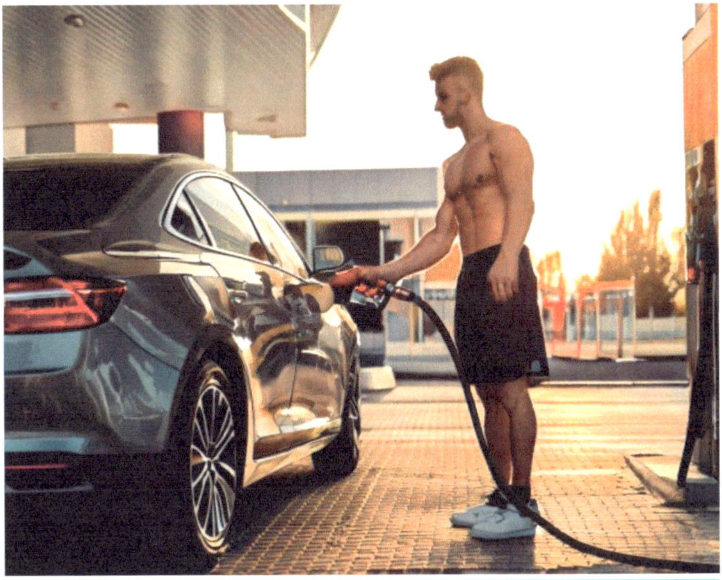

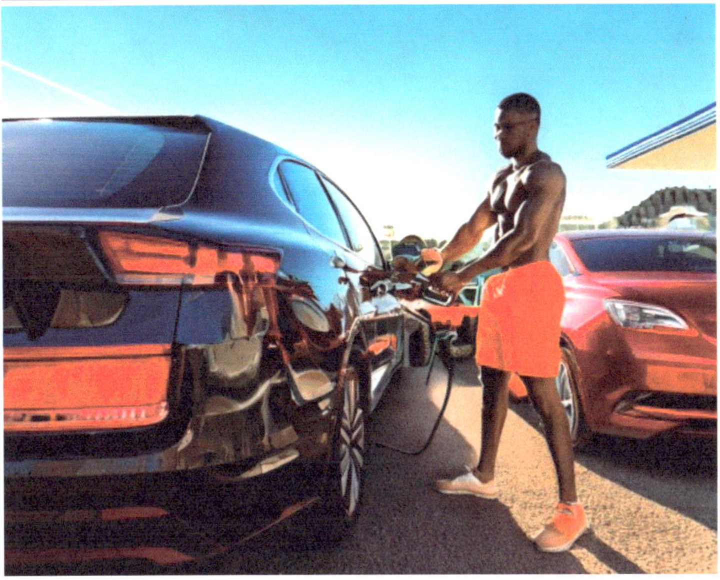

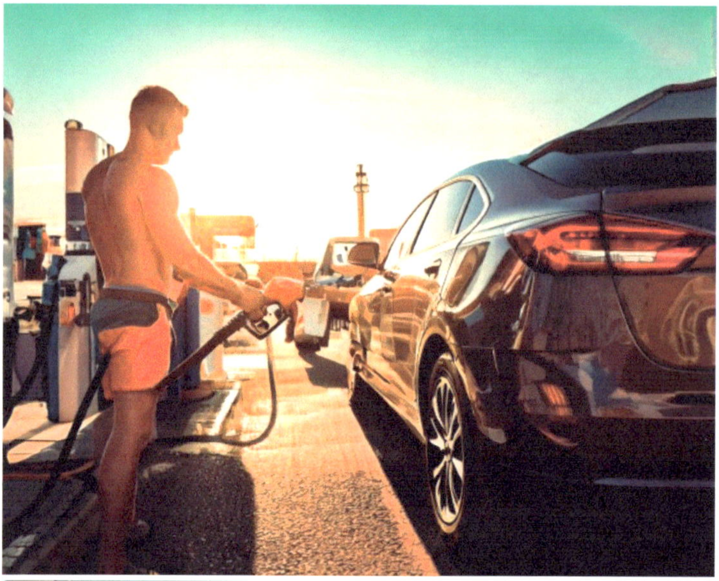

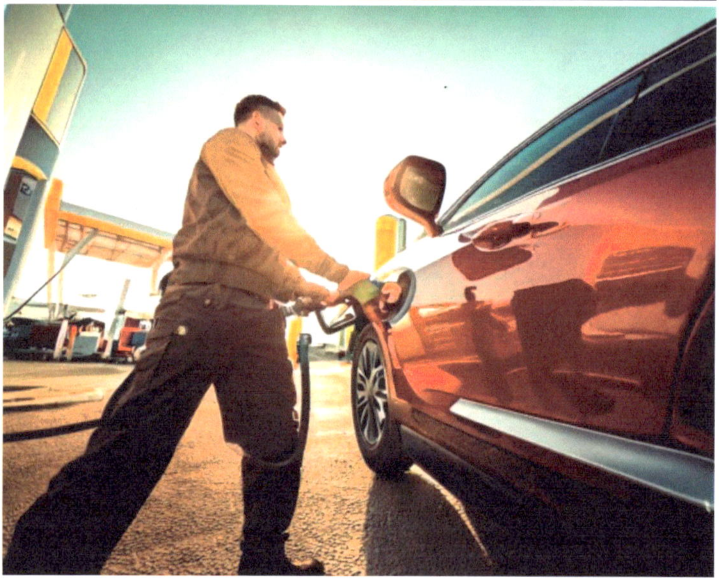

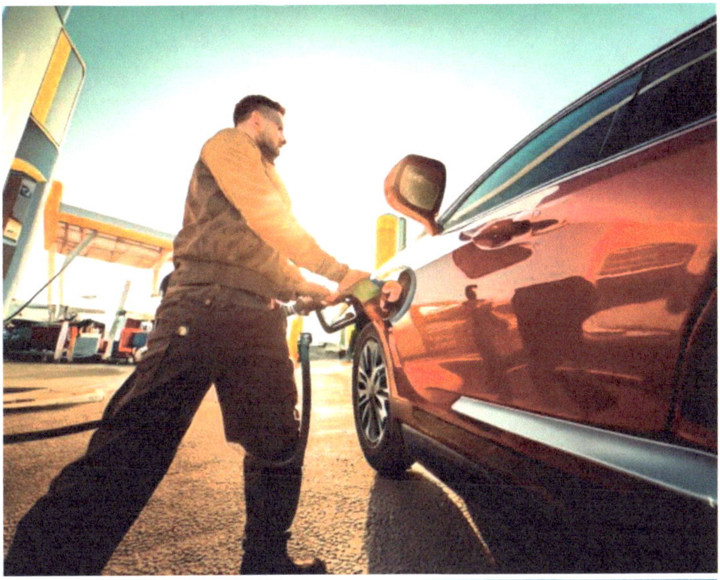

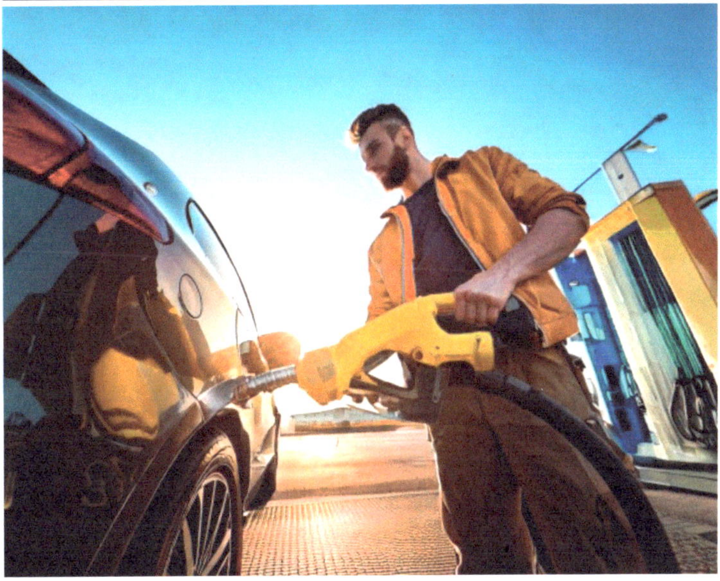

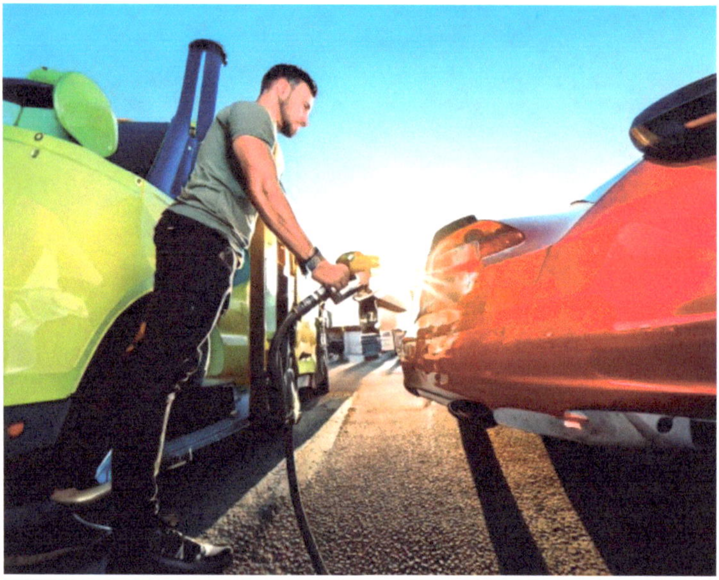

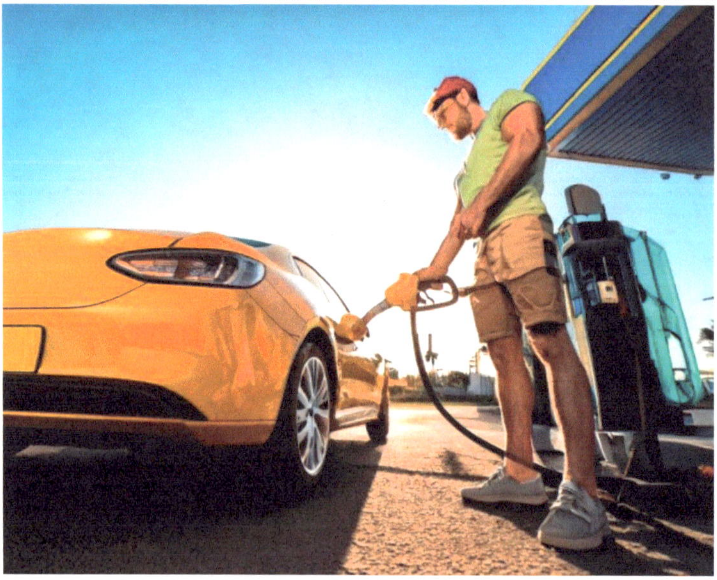

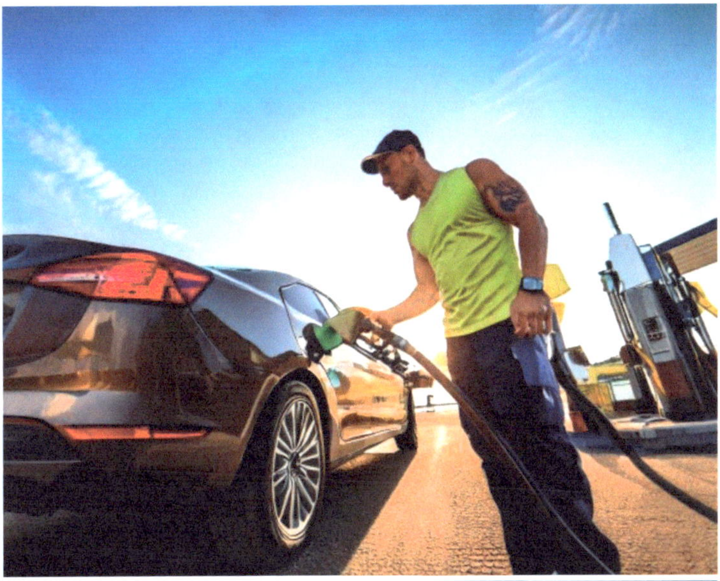

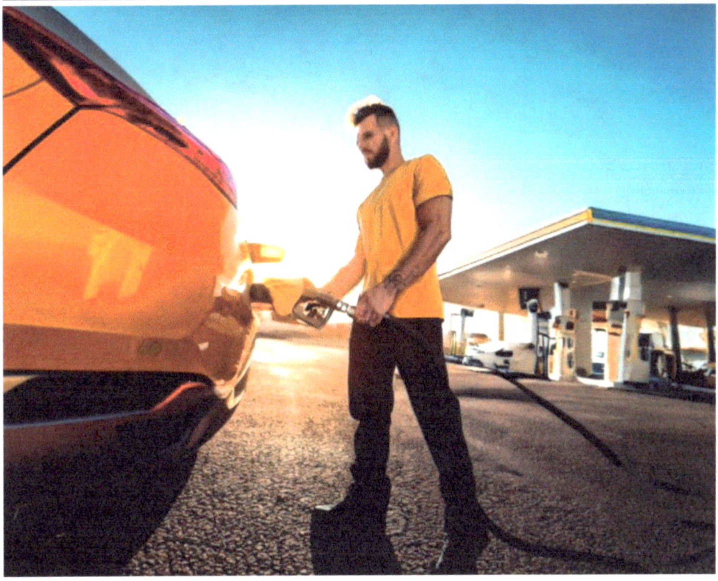

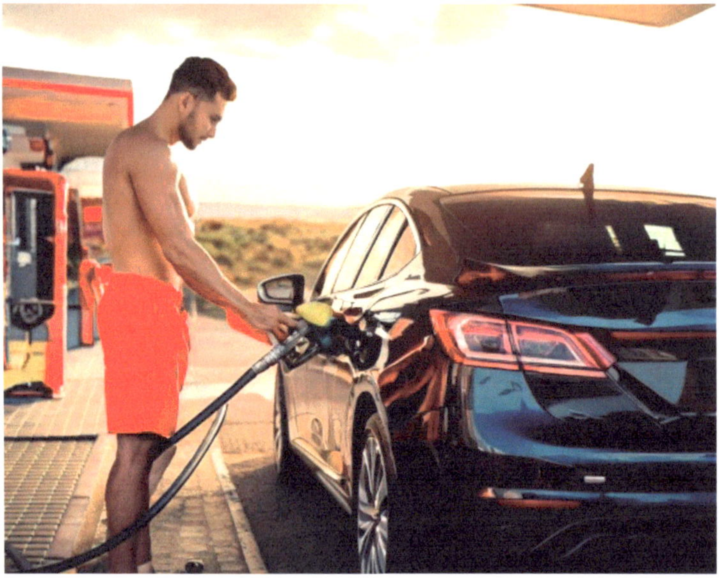

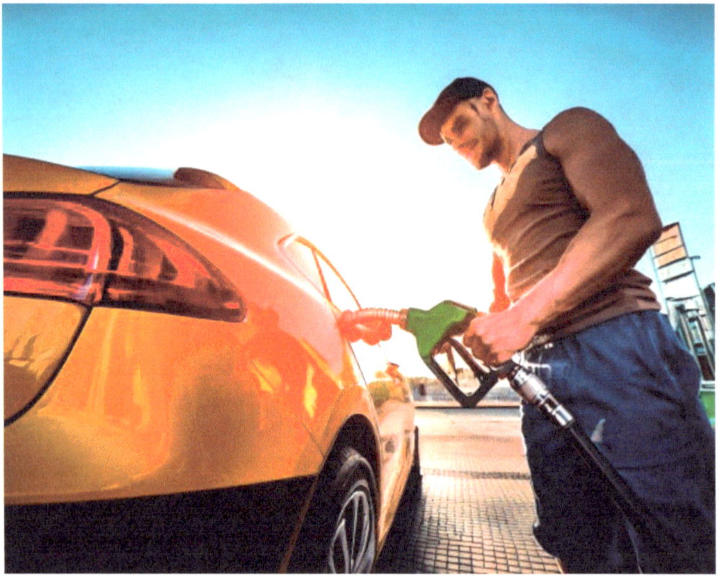

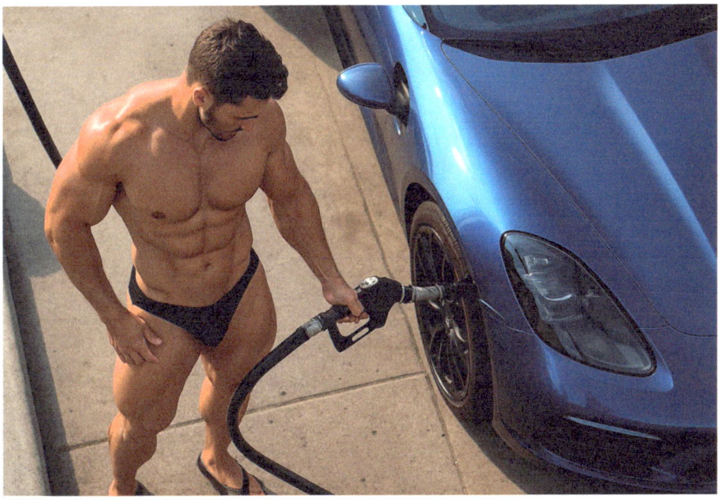

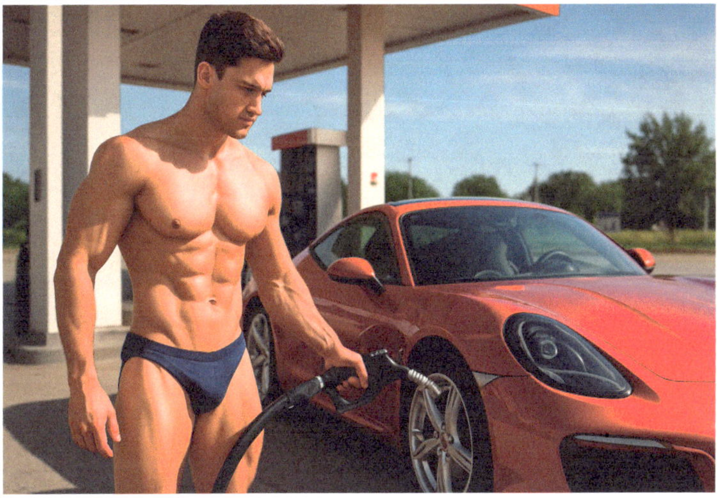

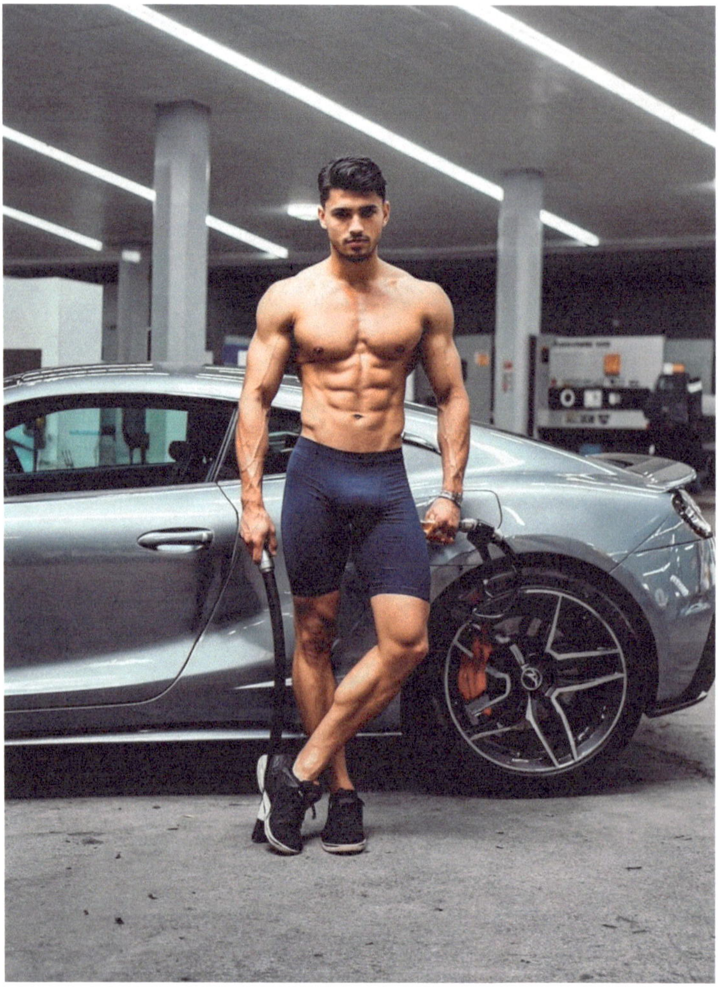

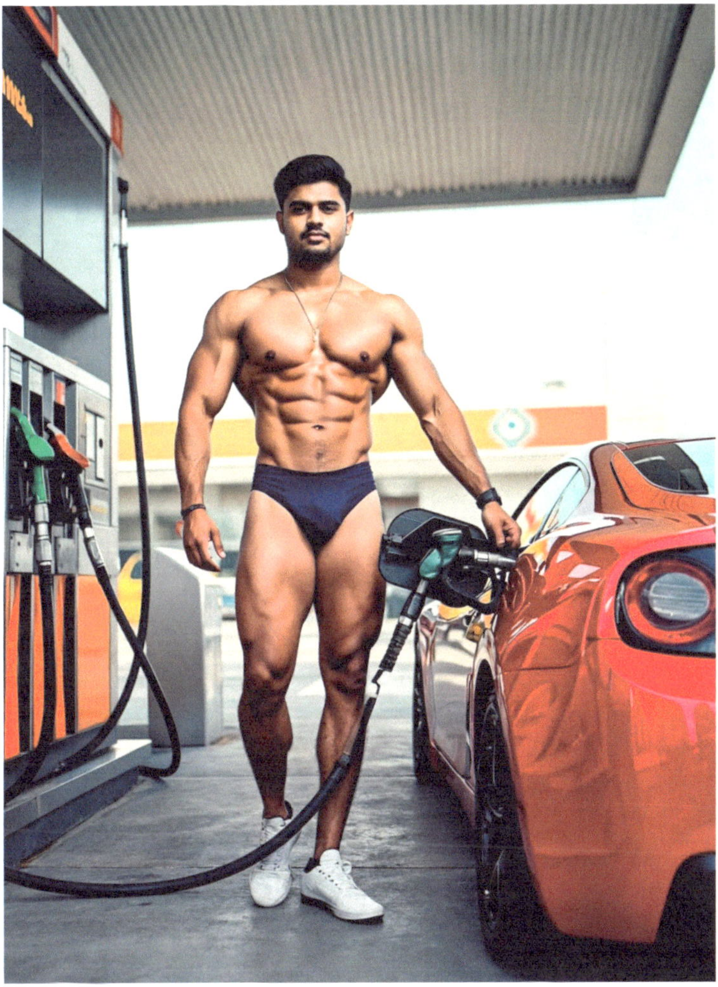

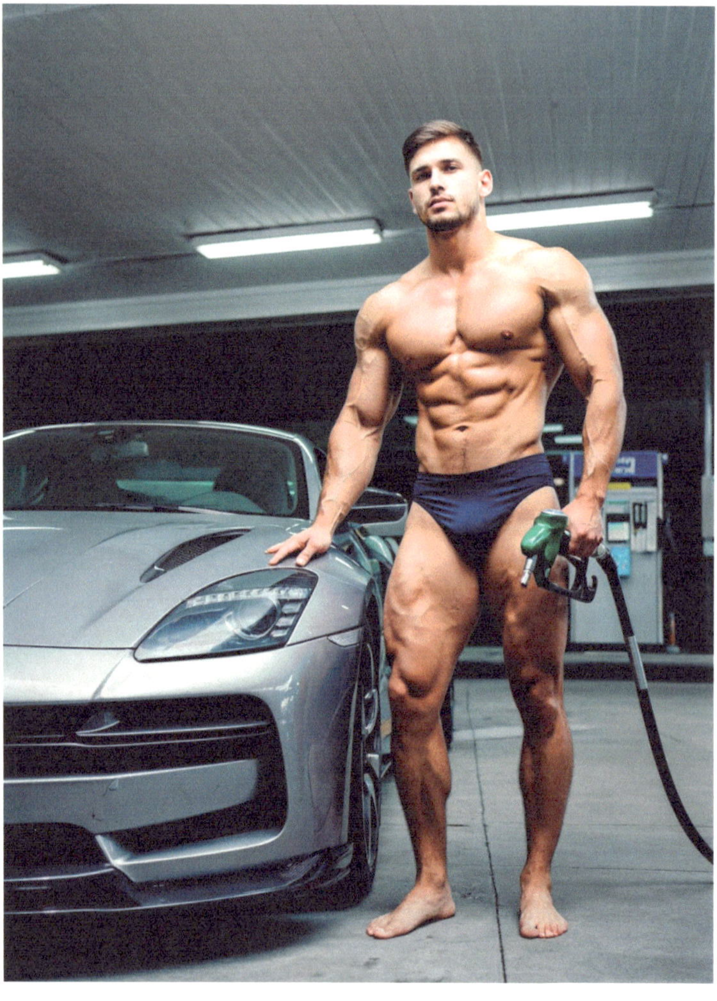

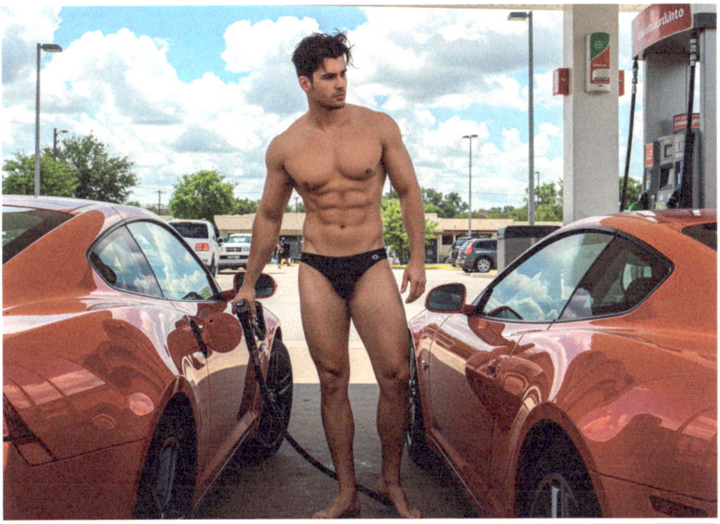

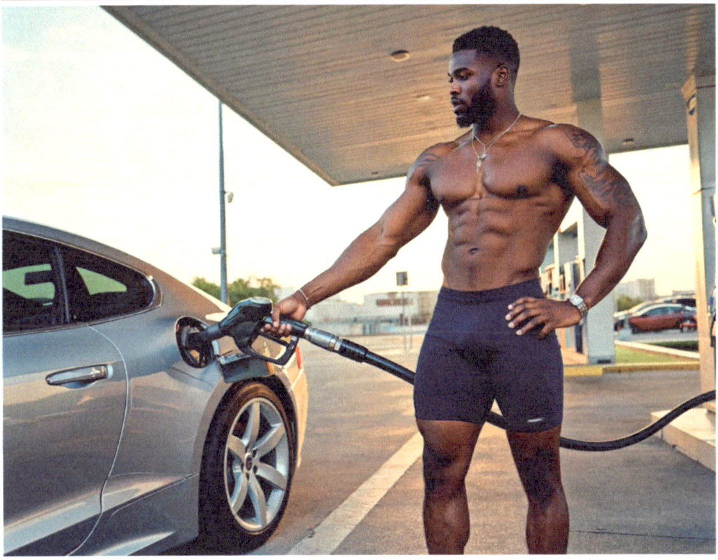

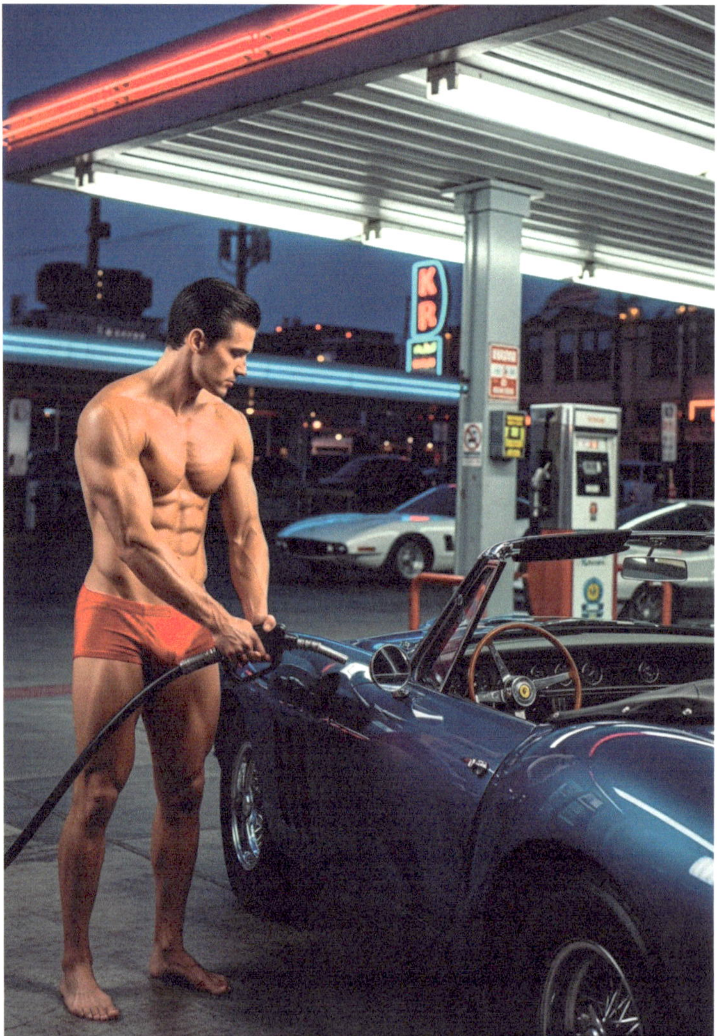
KR

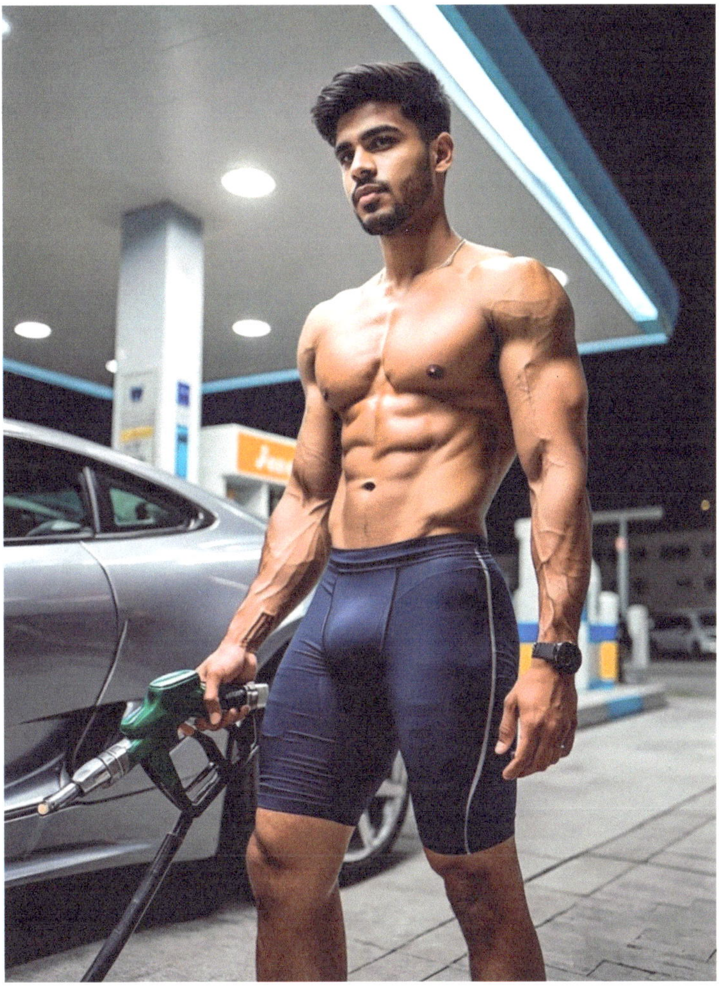

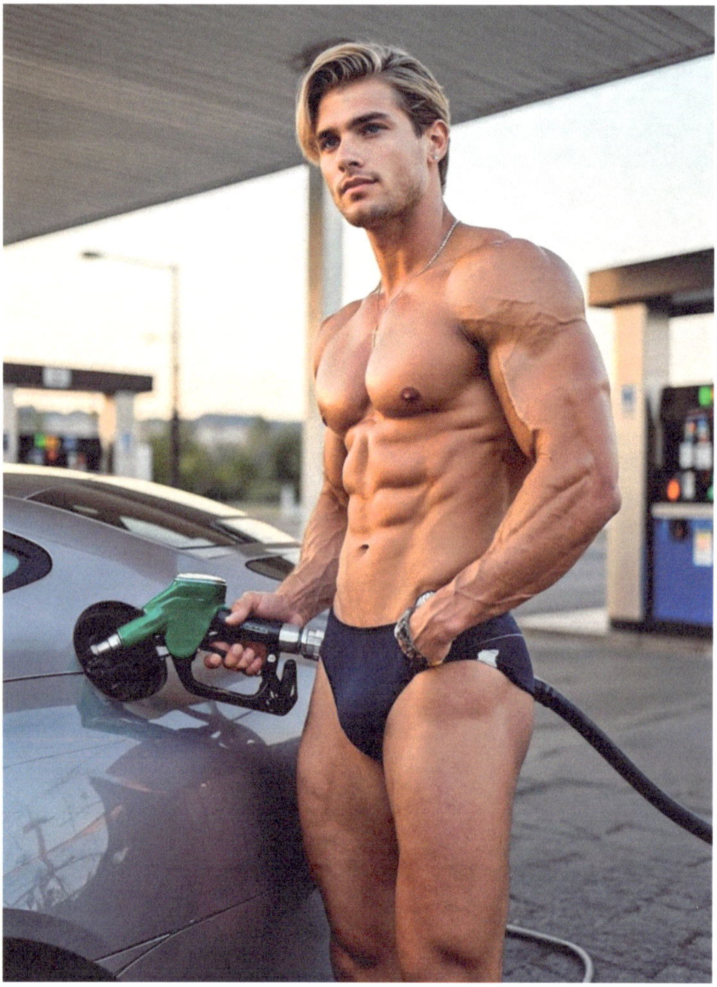

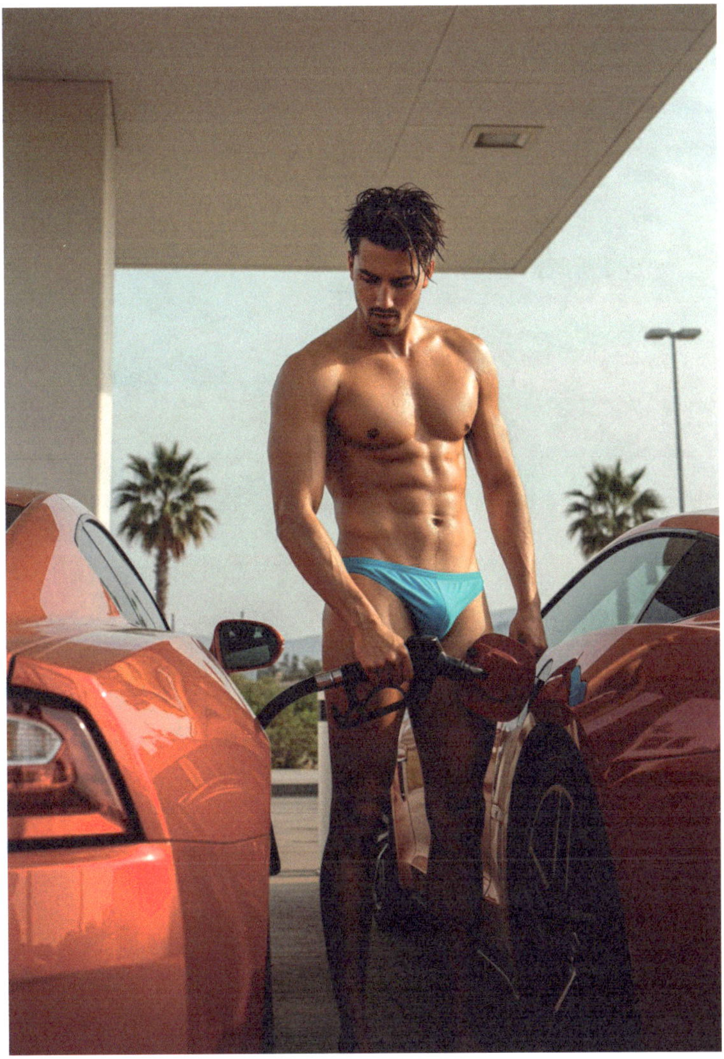

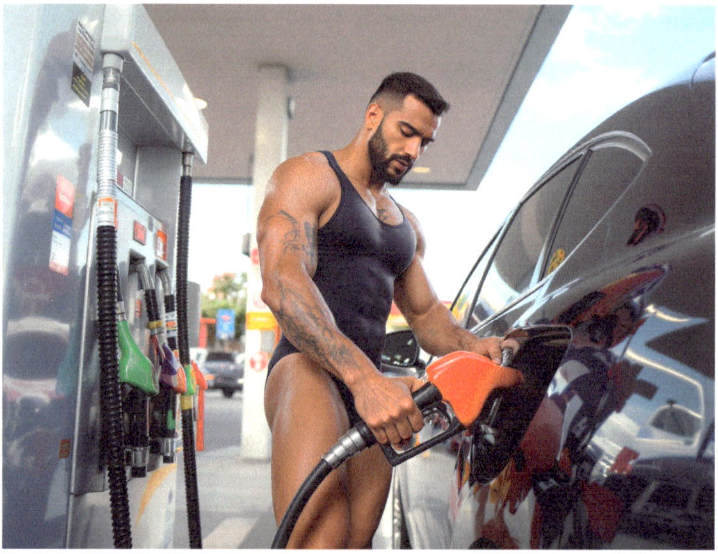

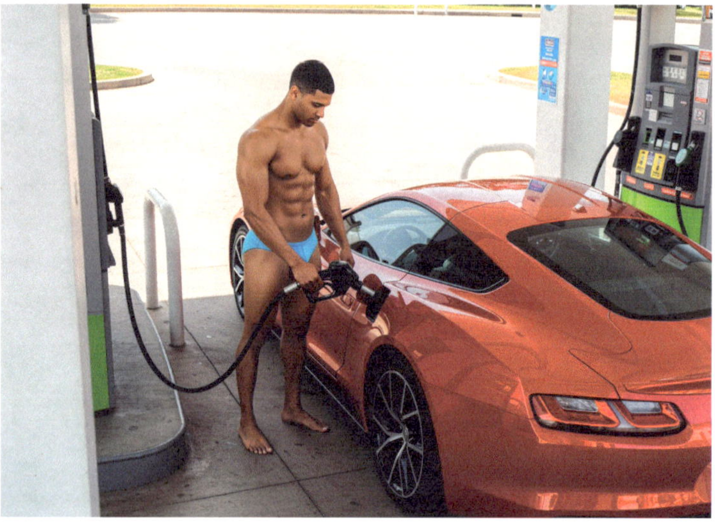

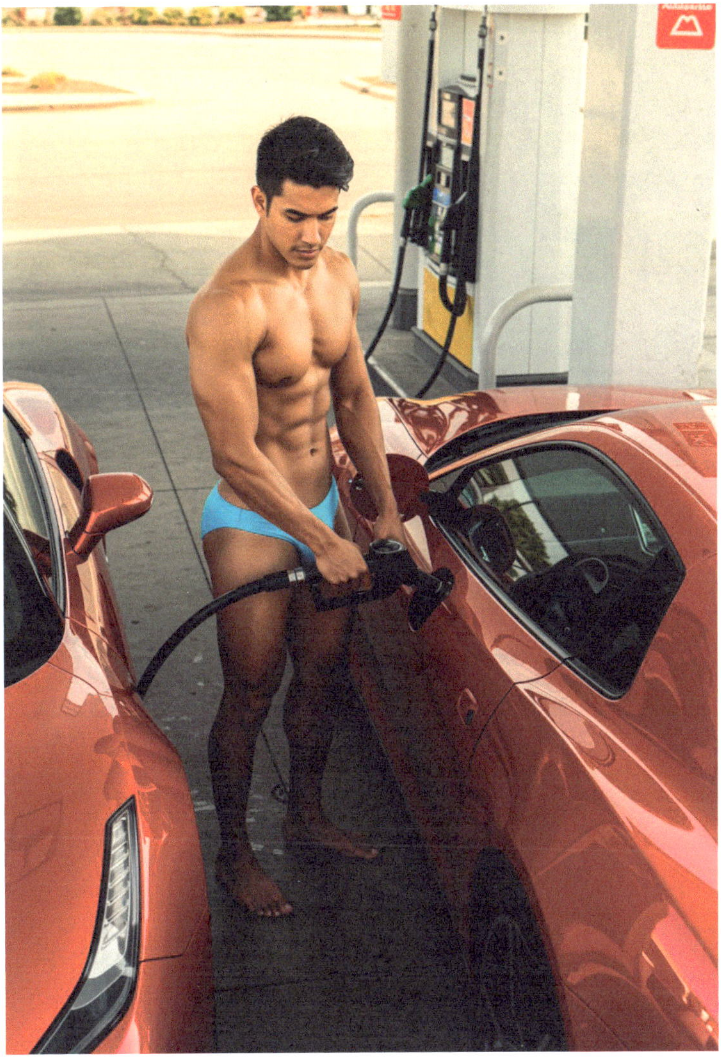

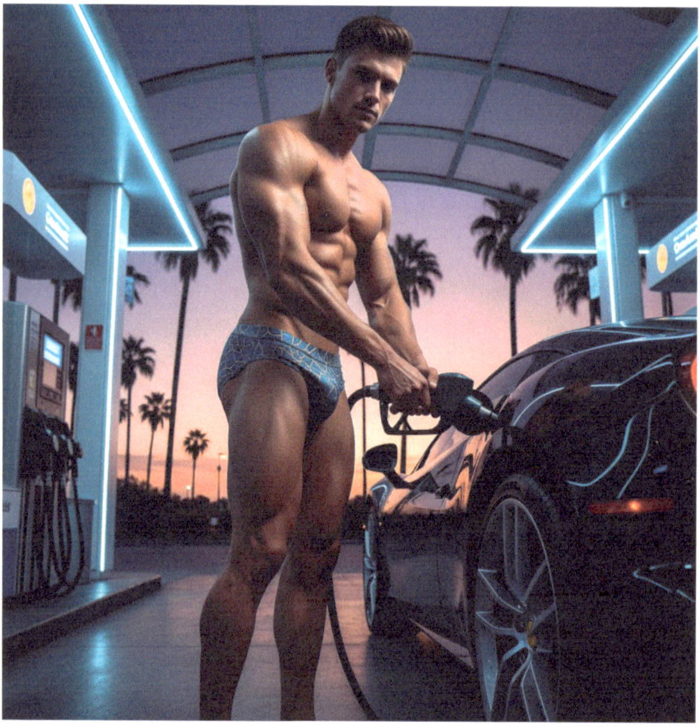

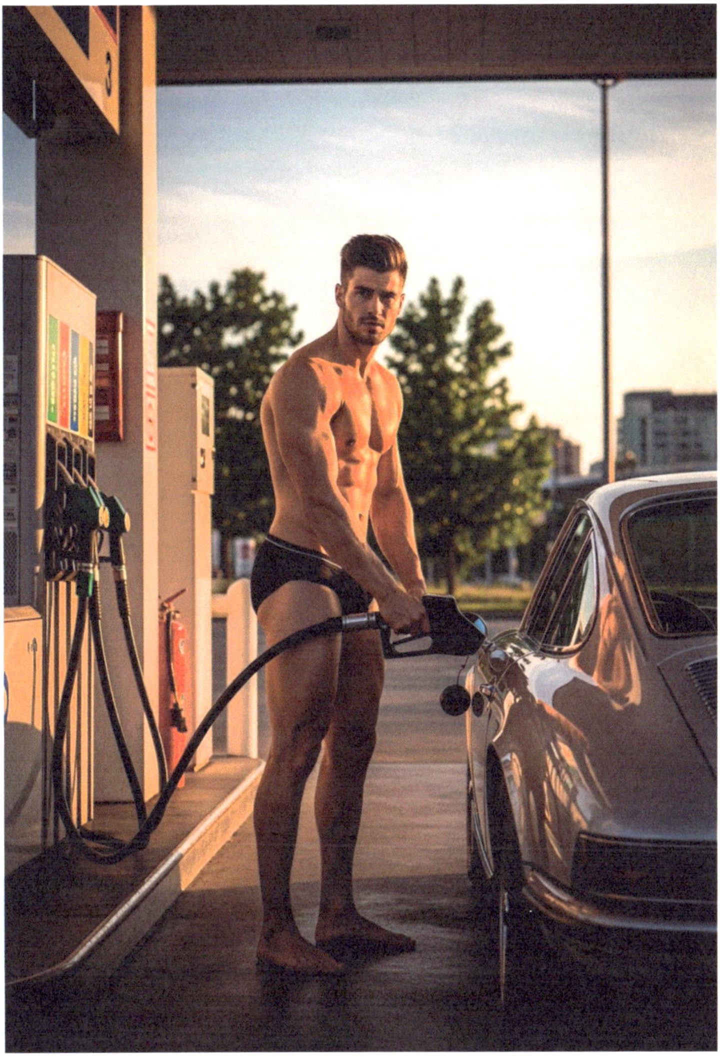

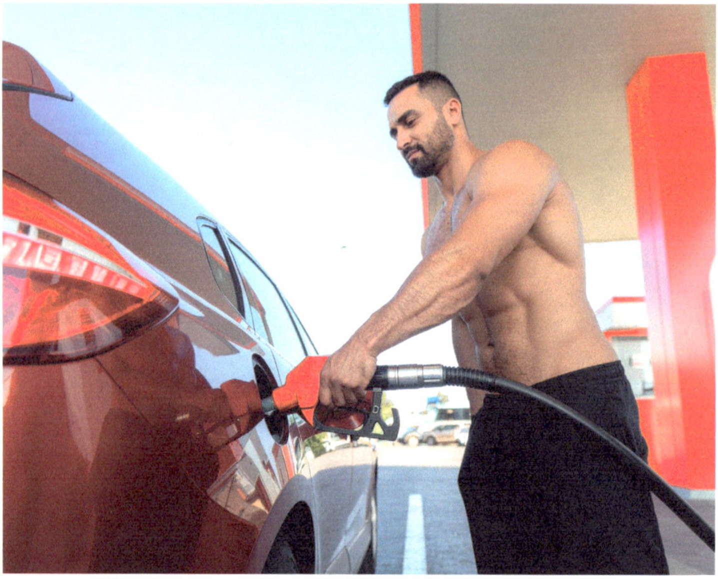

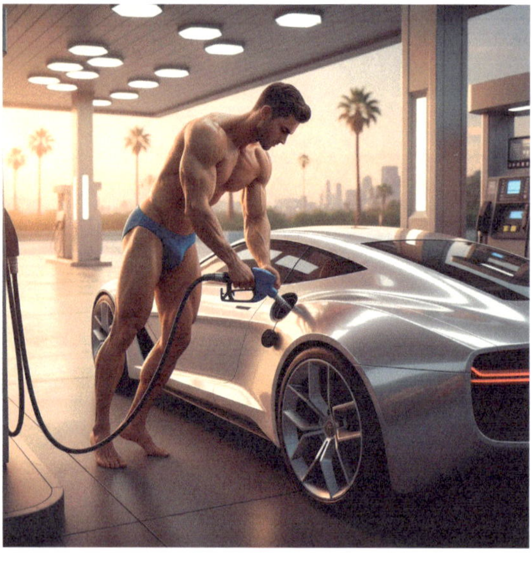

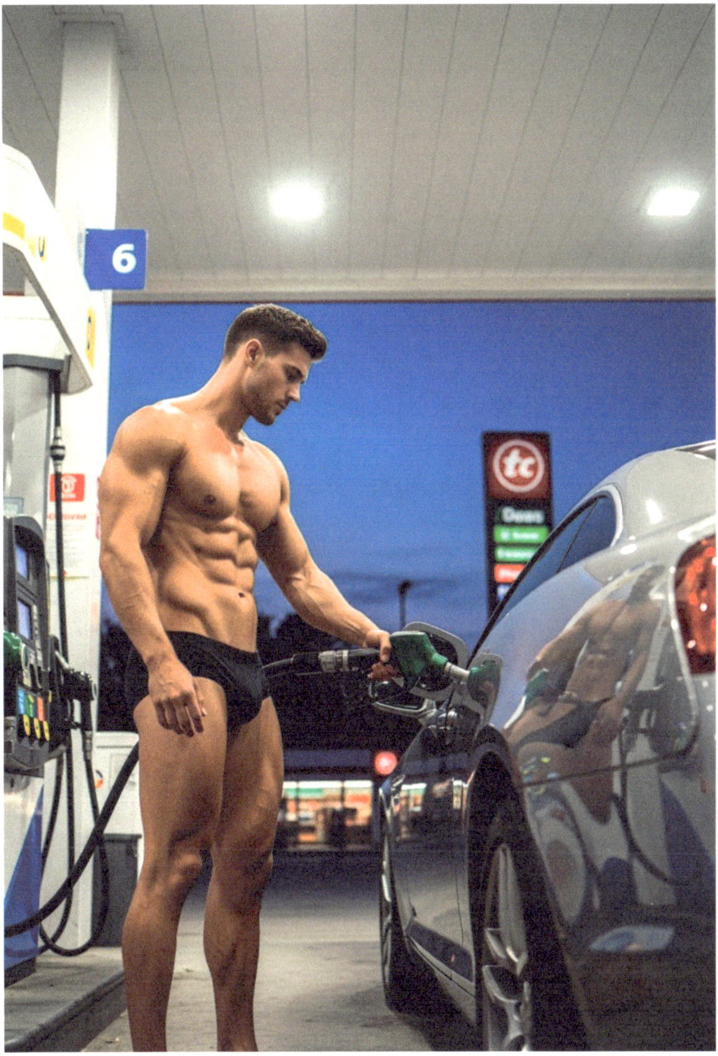
6
tc

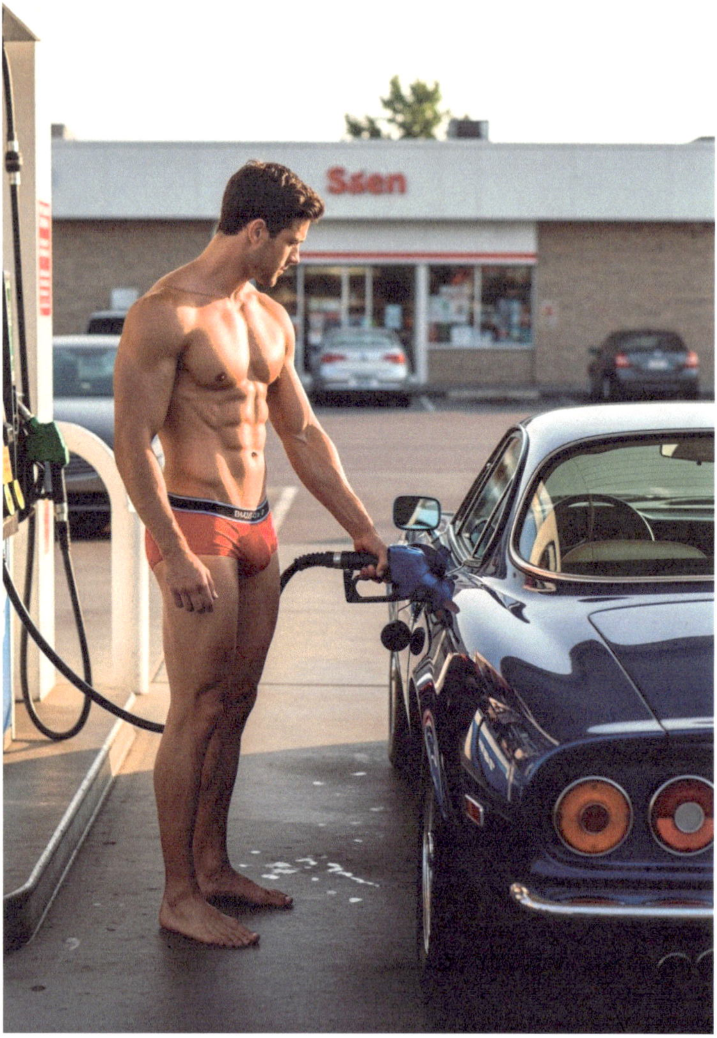

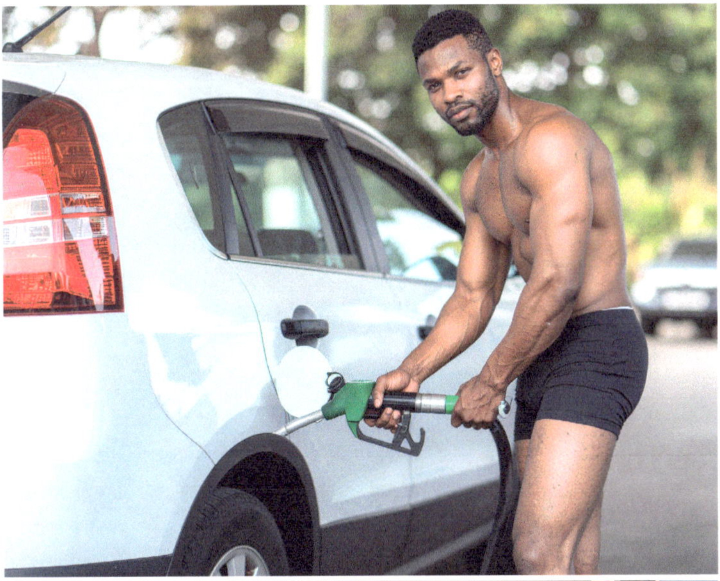

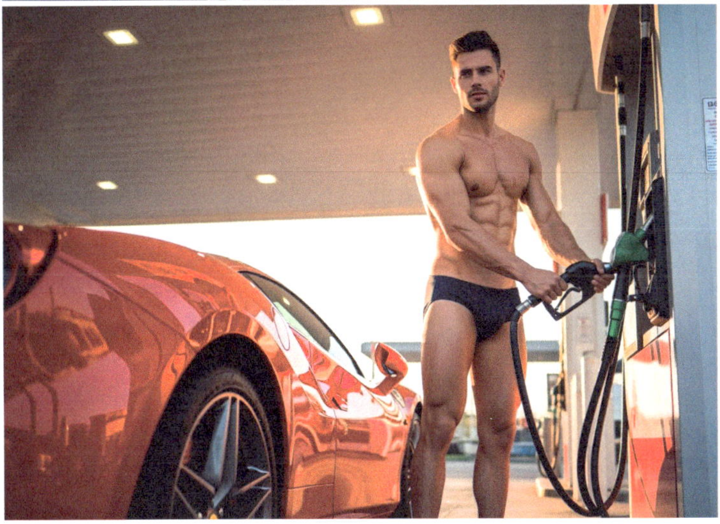

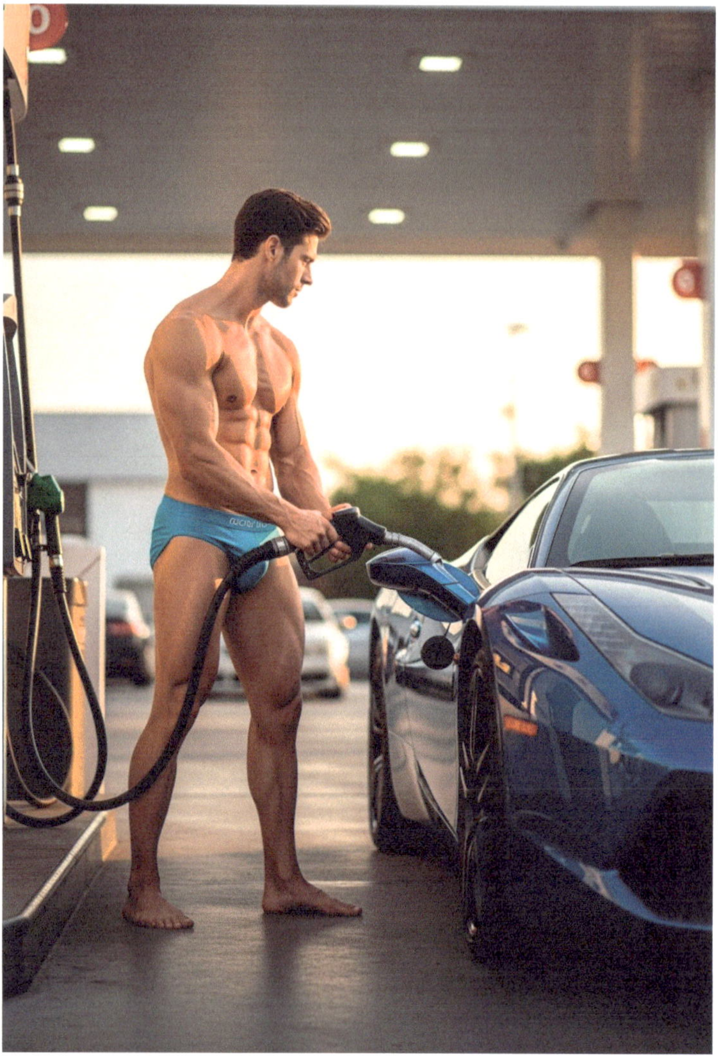

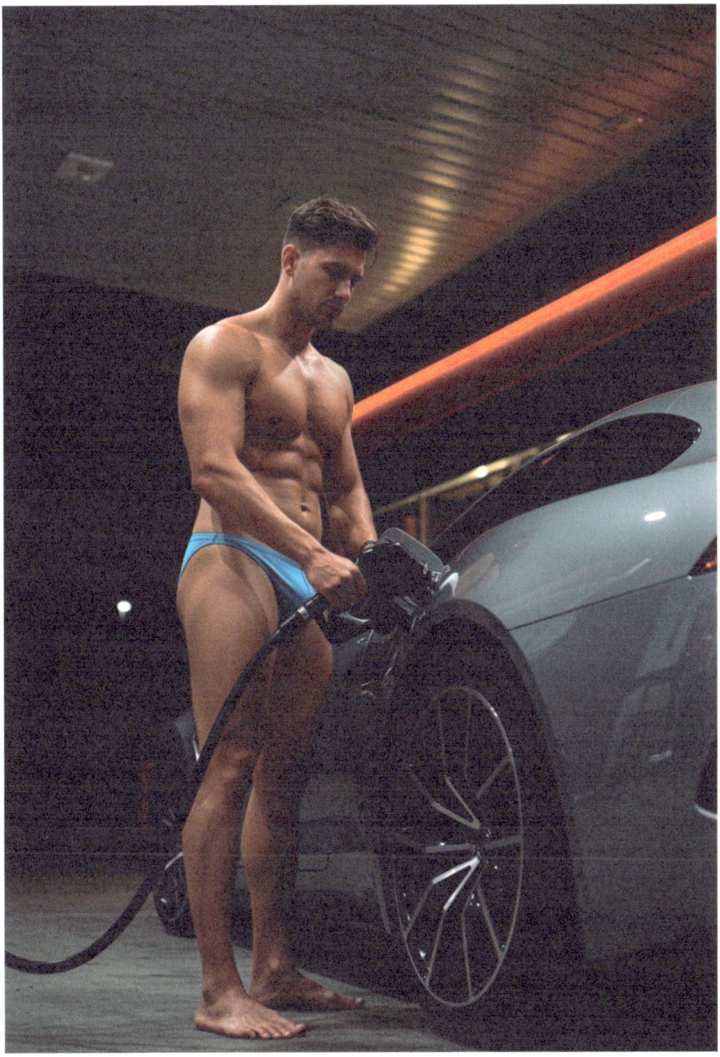

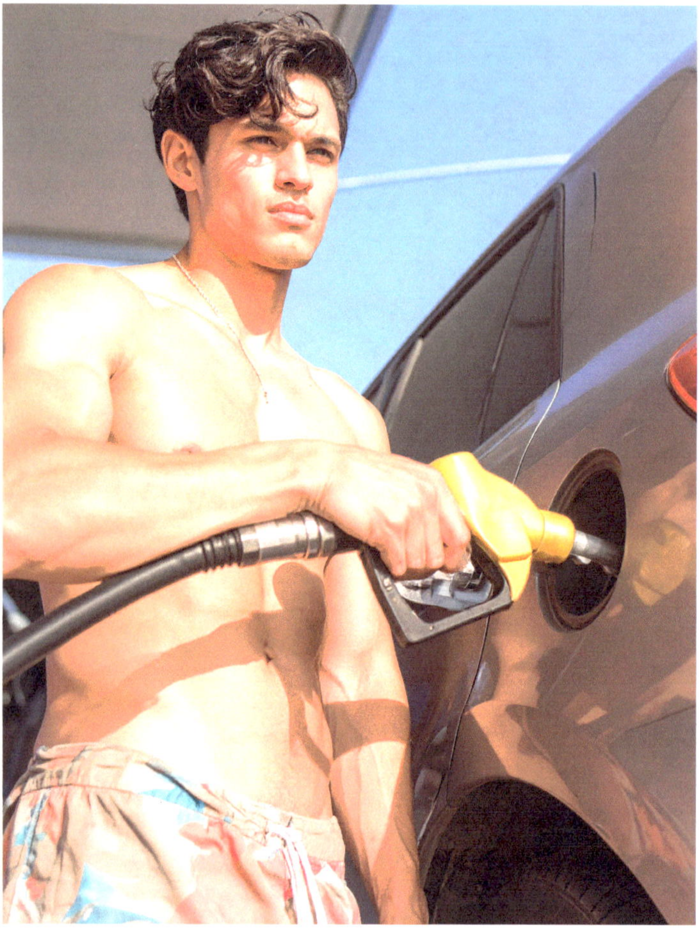

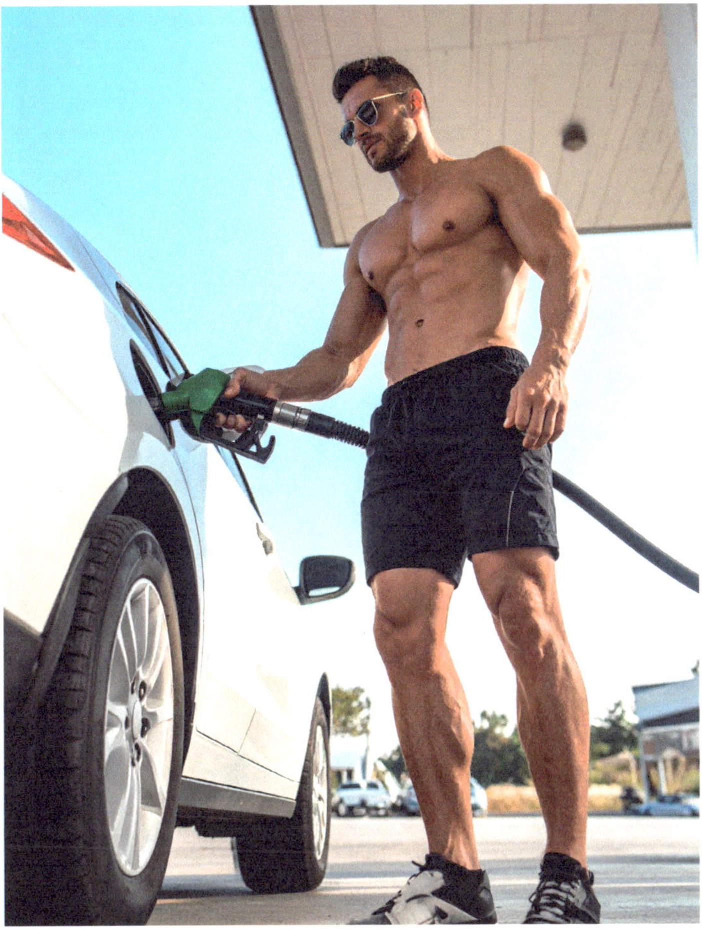

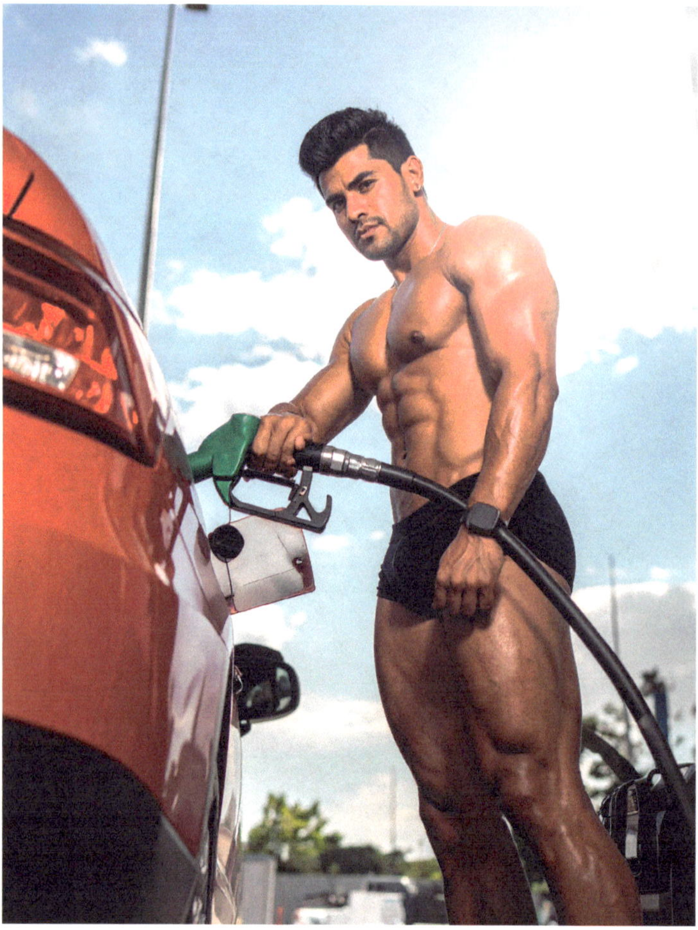

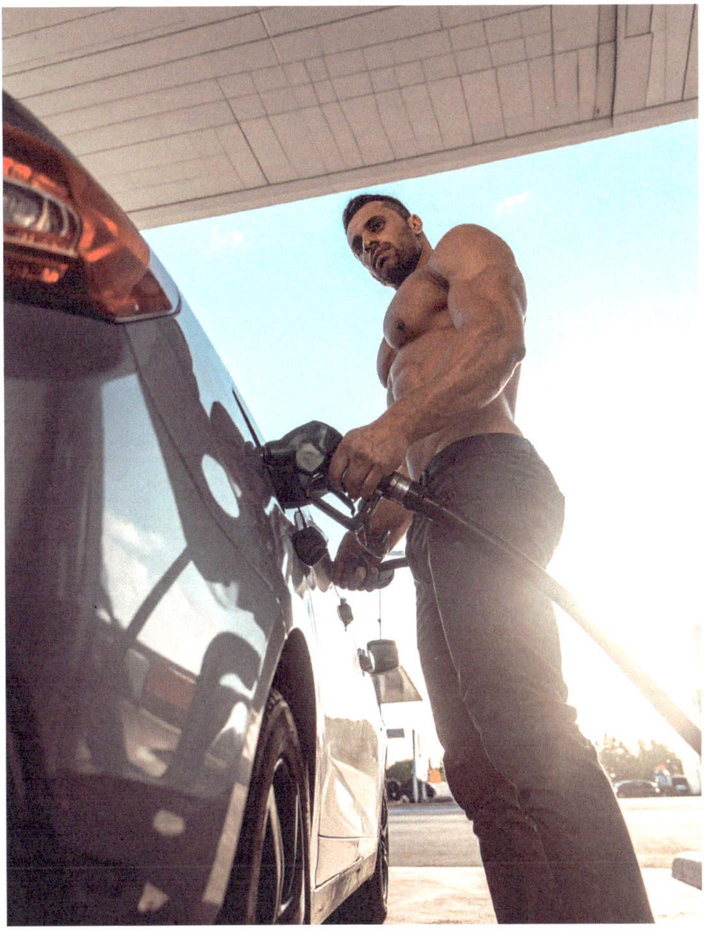

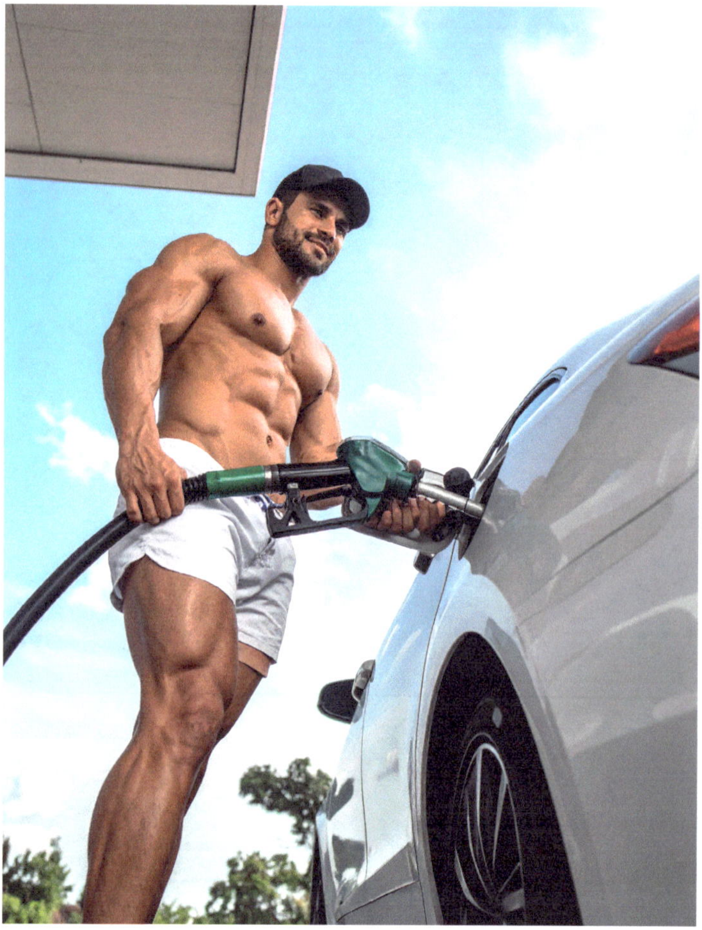

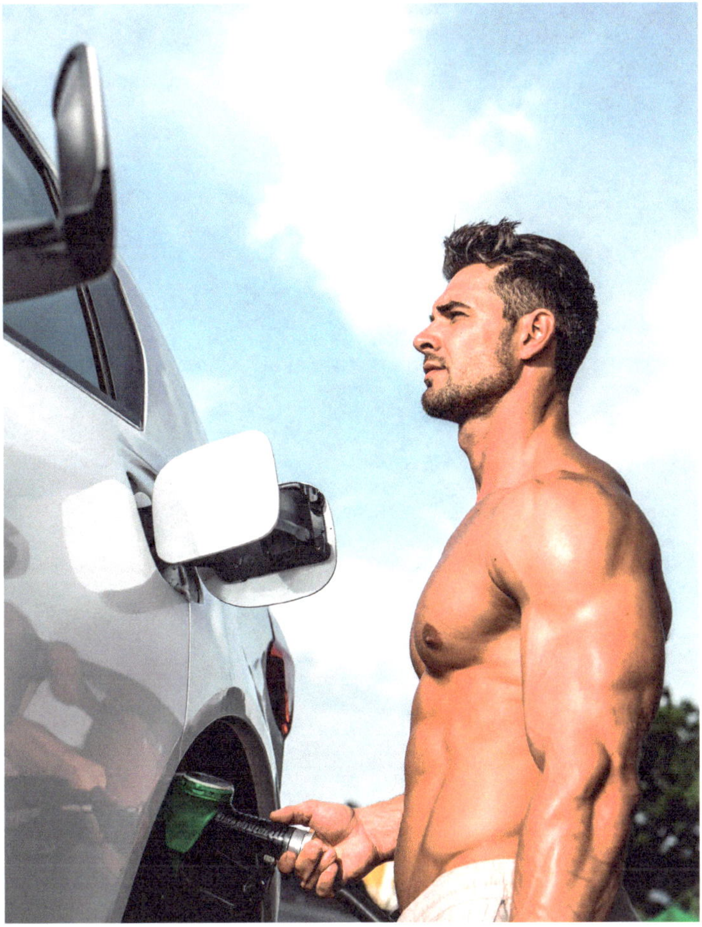

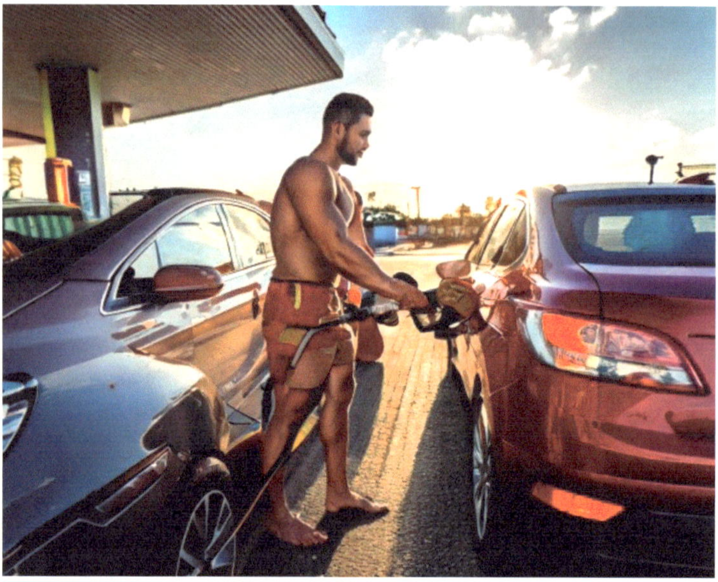

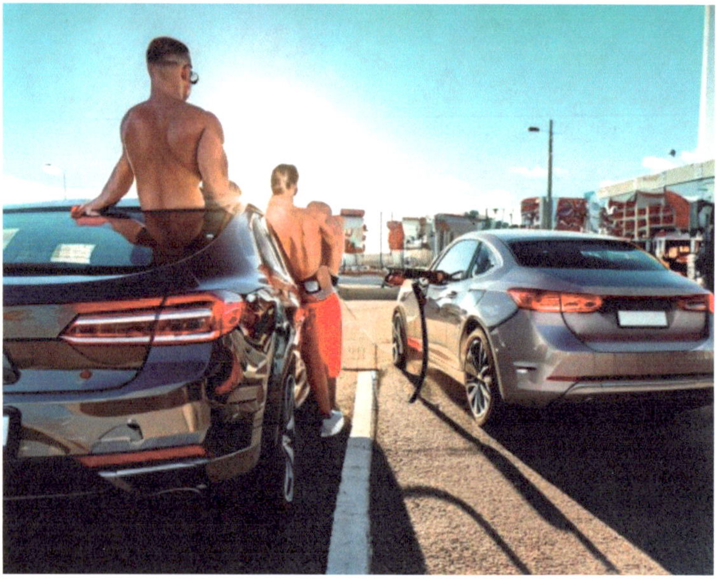

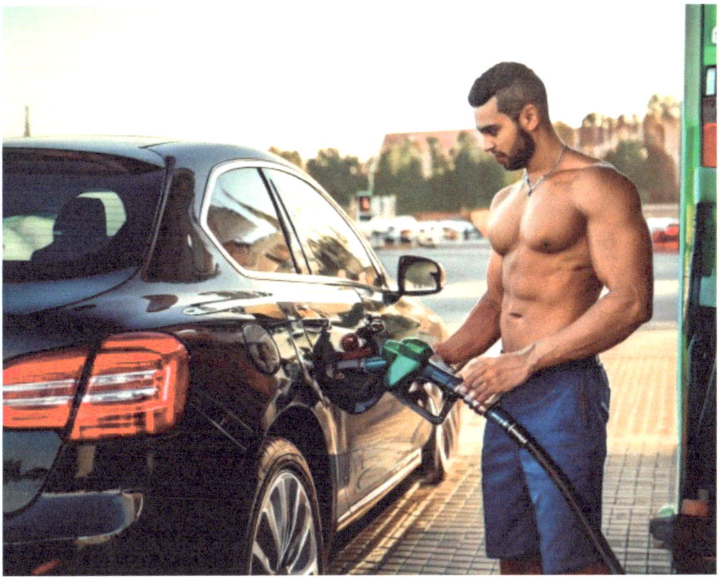

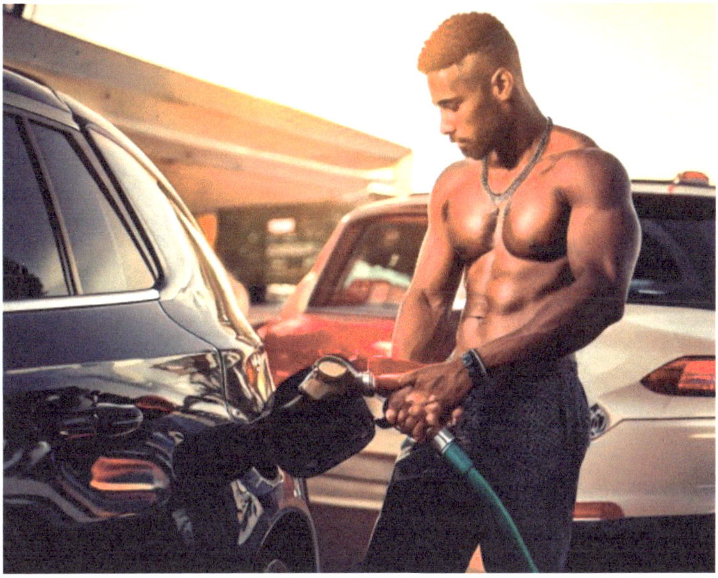

man in only chaps pumping gas into a car at a gas station

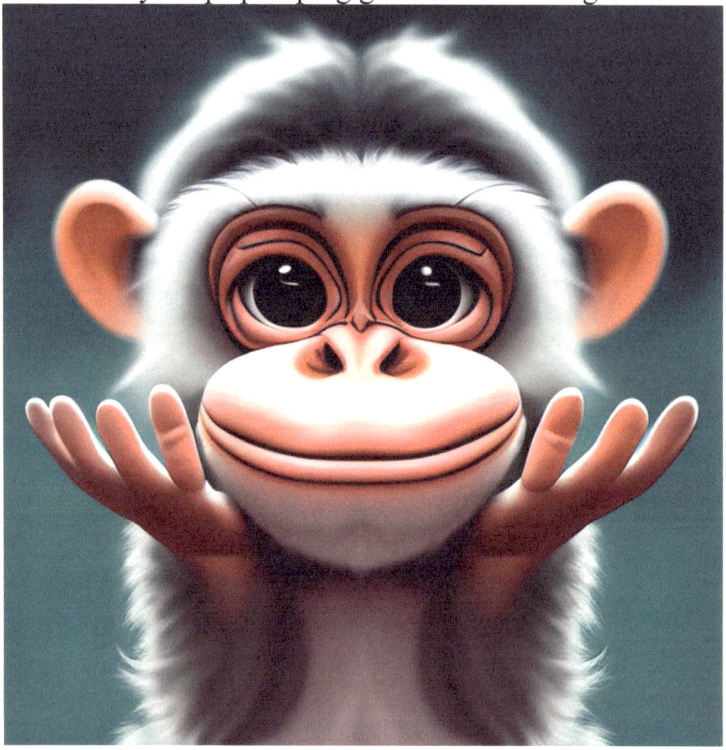

Can't load

We can't process this prompt. You can edit and try again, or flag for review. Note that prompt data is collected if flagged for review.

man in only chaps pumping gas into a car at a gas station

man in only chaps pumping gas into a car at a gas station shot from behind

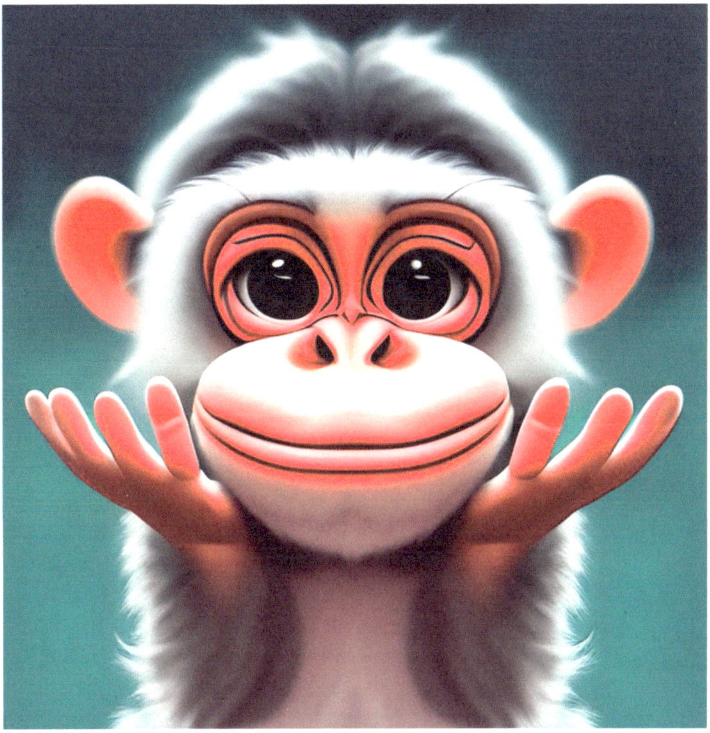

Can't load

We can't display the generated image.
You can edit and try again.

man in only chaps pumping gas into a car at a gas station shot from behind

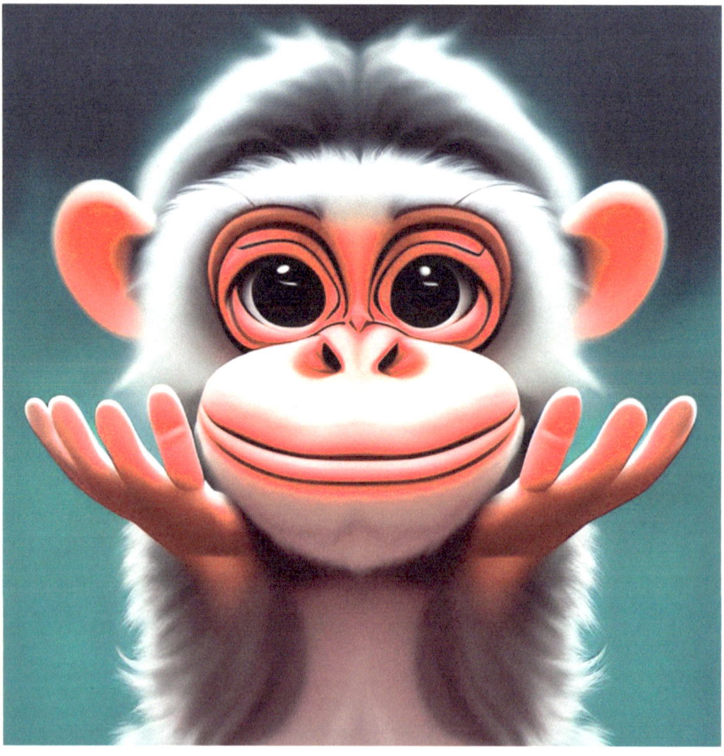

Can't load

We can't display the generated image.
You can edit and try again.

man in only chaps pumping gas into a car at a gas station shot from behind

man in only chaps pumping gas into a car at a gas station shot from behind

man in only chaps pumping gas into a car at a gas station shot from behind so we can see glutes

man in only chaps pumping gas into a car at a gas station shot from behind so we can see glutes

Something went wrong

Please try again later.

man in only chaps pumping gas into a car at a gas station
shot from behind so we can see glutes

Something went wrong

Please try again later.

man in only chaps pumping gas into a car at a gas station
shot from behind so we can see glutes

man in only chaps pumping gas into a car at a gas station
shot from behind so we can see glutes

man in only chaps pumping gas into a car at a gas station
shot from behind so we can see glutes

man in only chaps pumping gas into a car at a gas station
shot from behind so we can see glutes

man in only chaps pumping gas into a car at a gas station
shot from behind so we can see glutes

man in only chaps pumping gas into a car at a gas station
shot from behind so we can see glutes

man in only chaps pumping gas into a car at a gas station shot from behind so we can see glutes

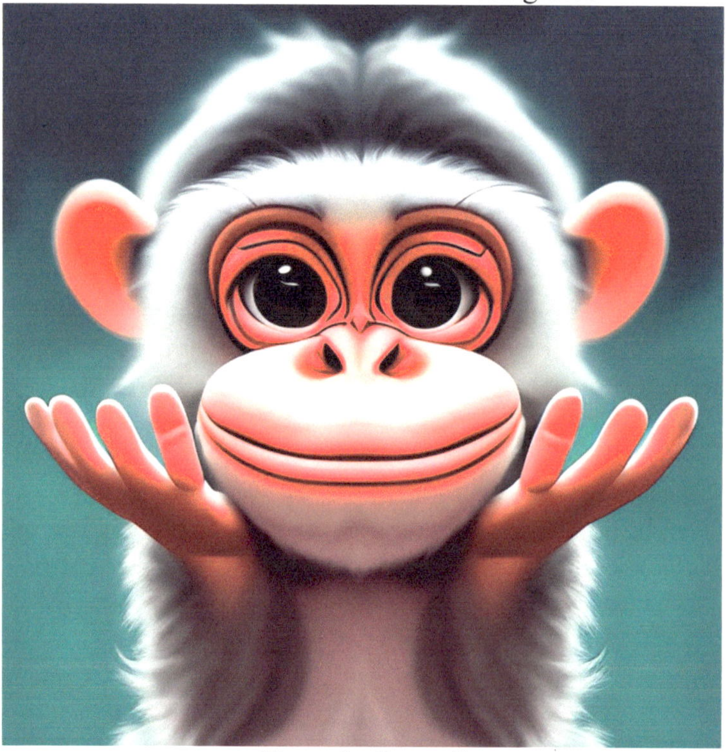

Can't load

We can't display the generated image.
You can edit and try again.

man in only chaps pumping gas into a car at a gas station shot from behind so we can see glutes

Something went wrong

Please try again later.

man in only a loincloth pumping gas into a car at a gas station

man in only a loincloth pumping gas into a car at a gas station

man in only a loincloth pumping gas into a car at a gas station

man in only a loincloth pumping gas into a car at a gas station

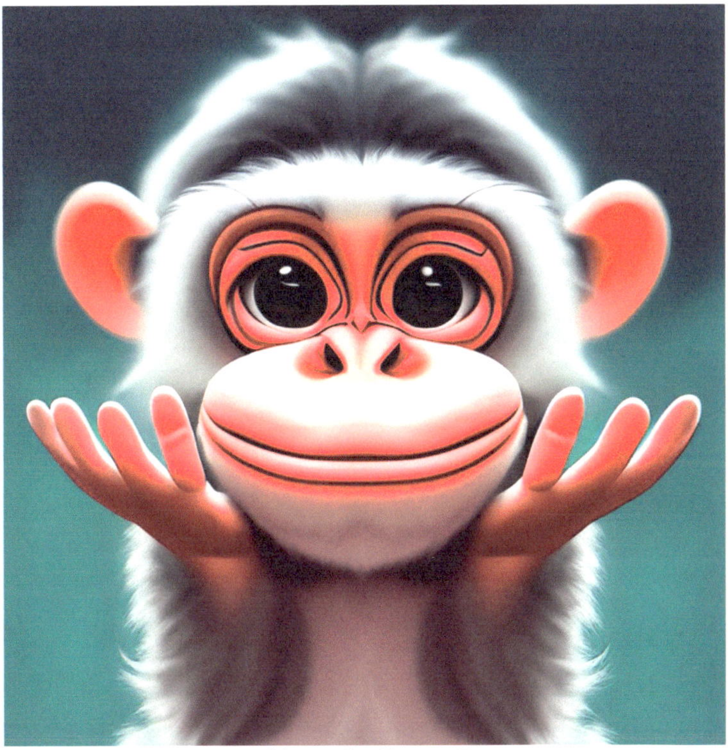

Can't load

We can't display the generated image.
You can edit and try again.

man in only a loincloth pumping gas into a car at a gas station

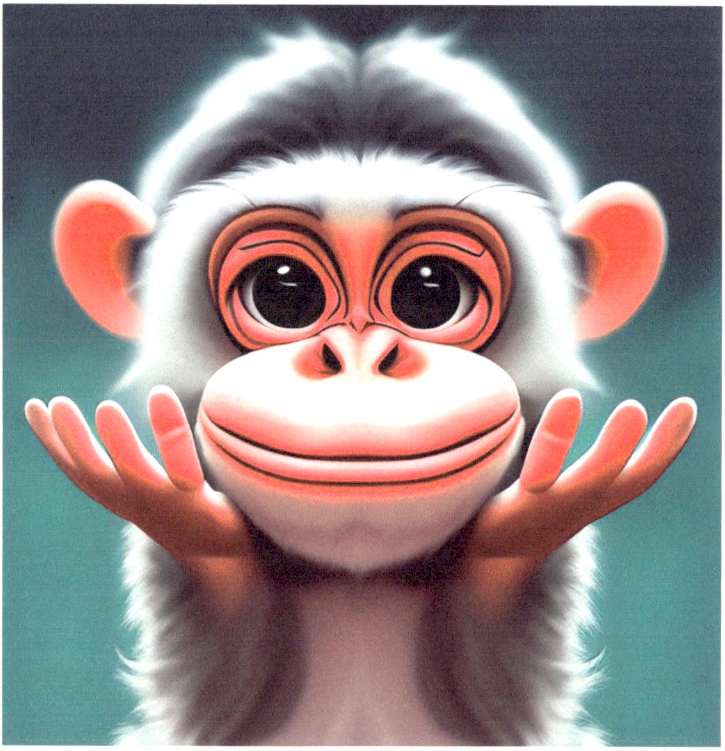

Can't load

We can't display the generated image.
You can edit and try again.

\

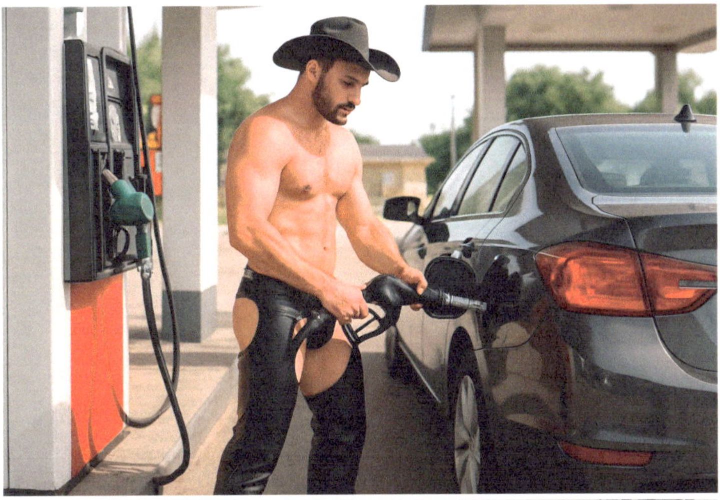

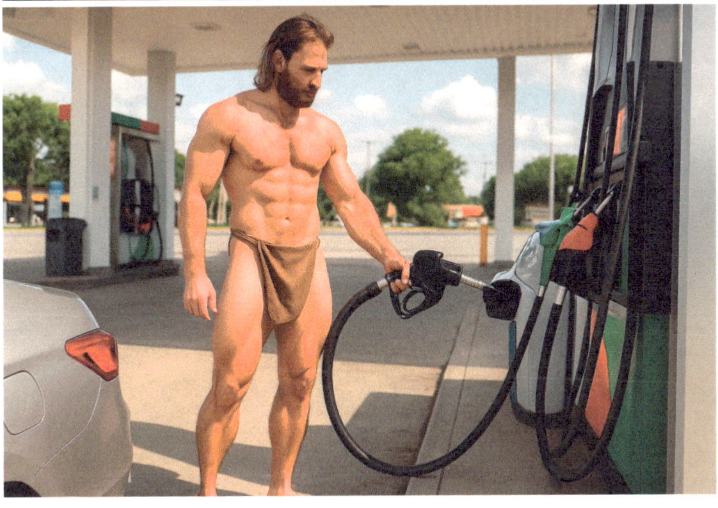

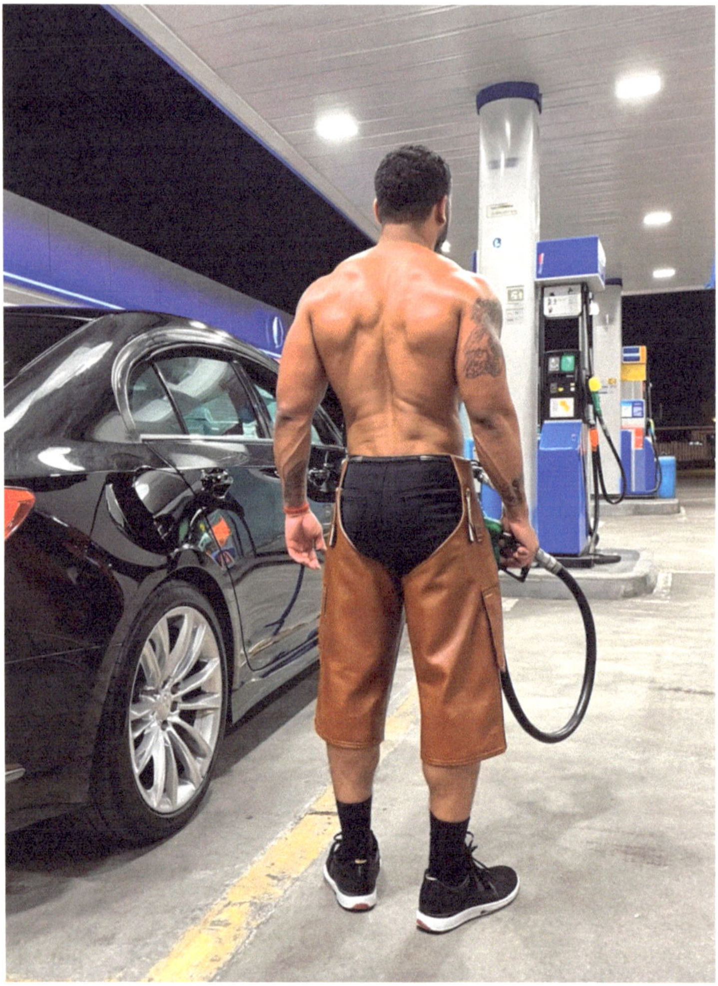

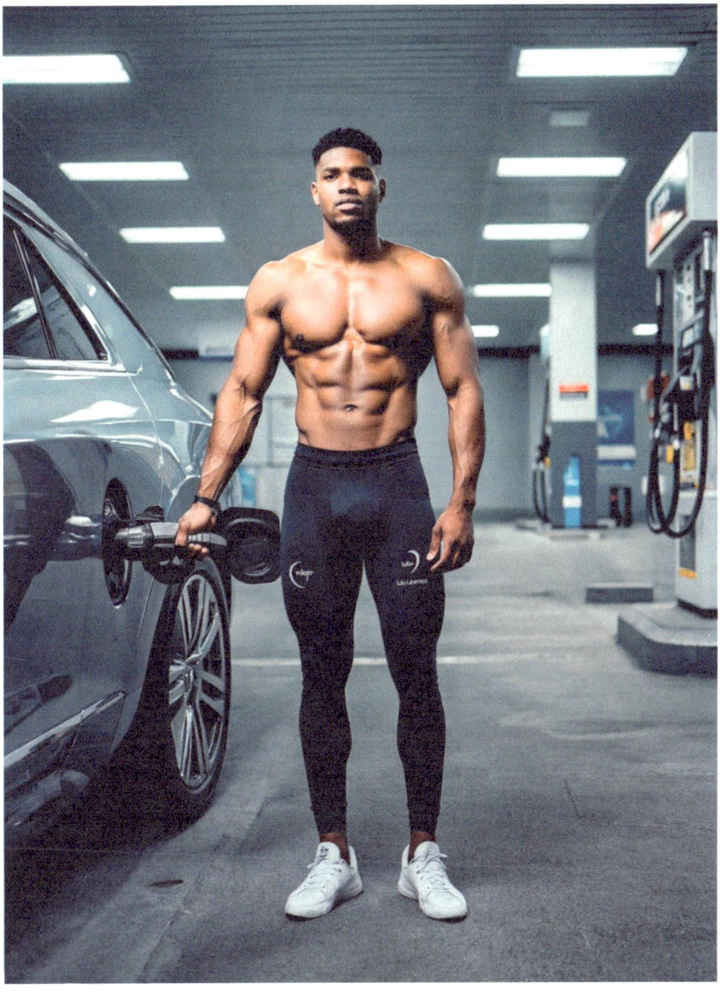

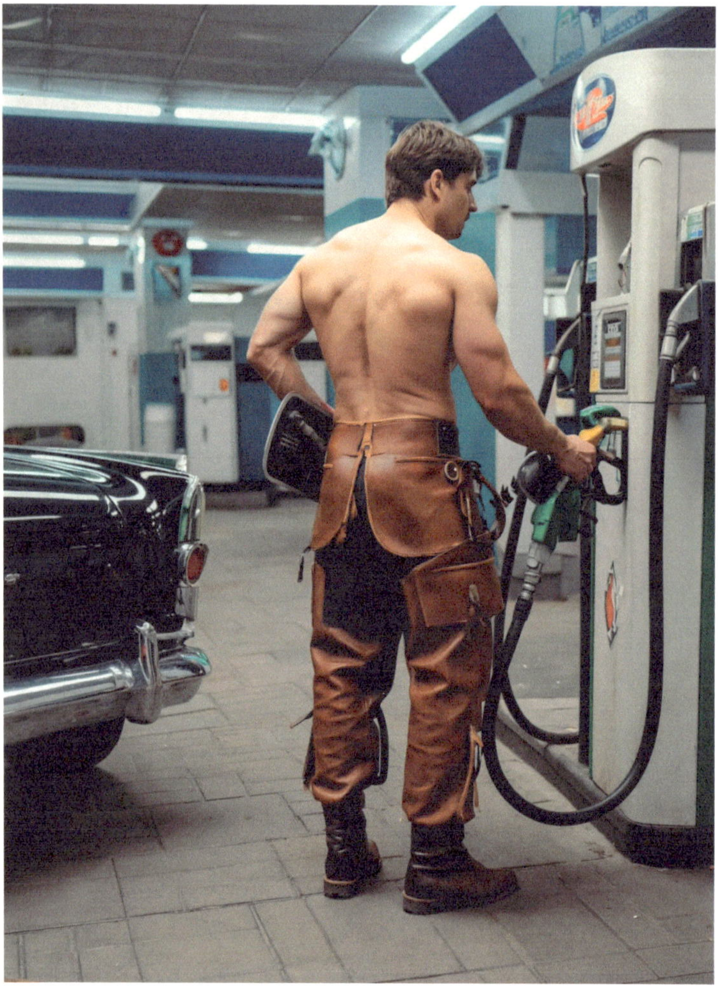

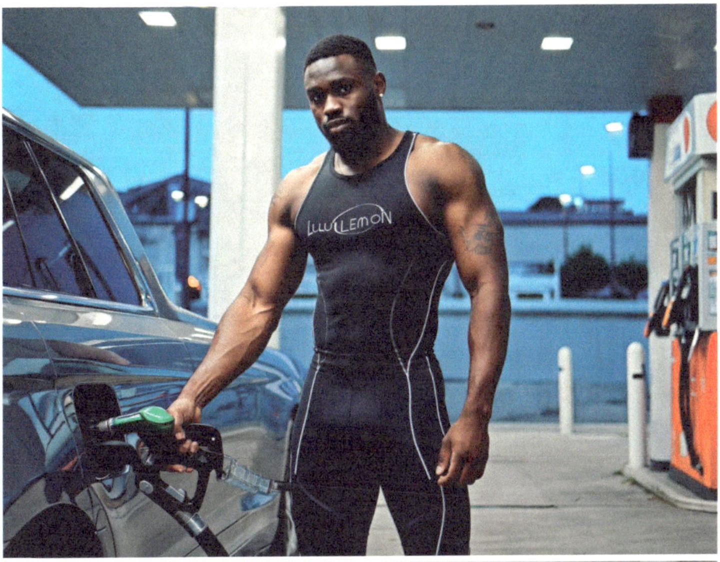
LULULEMON

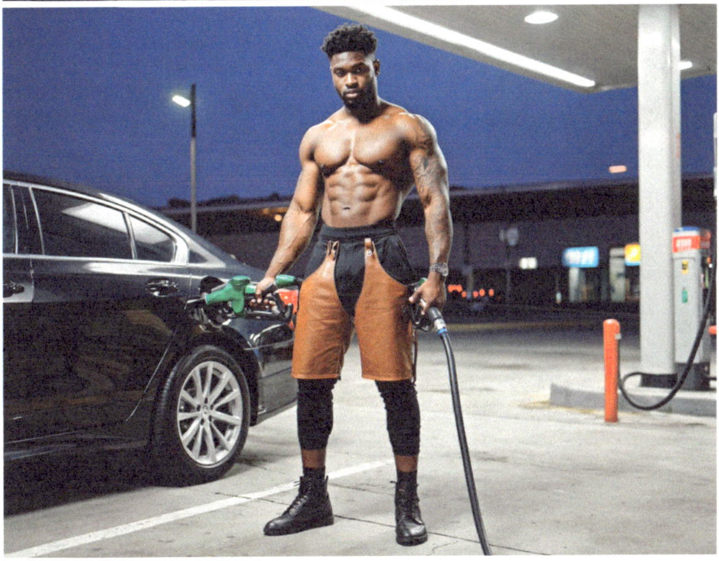

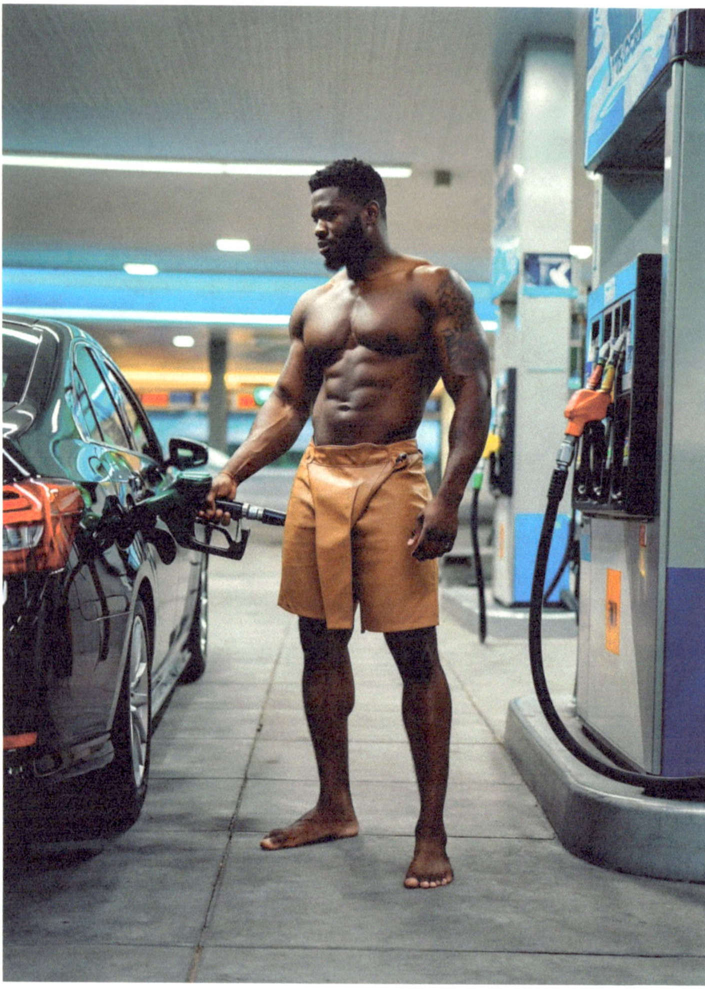

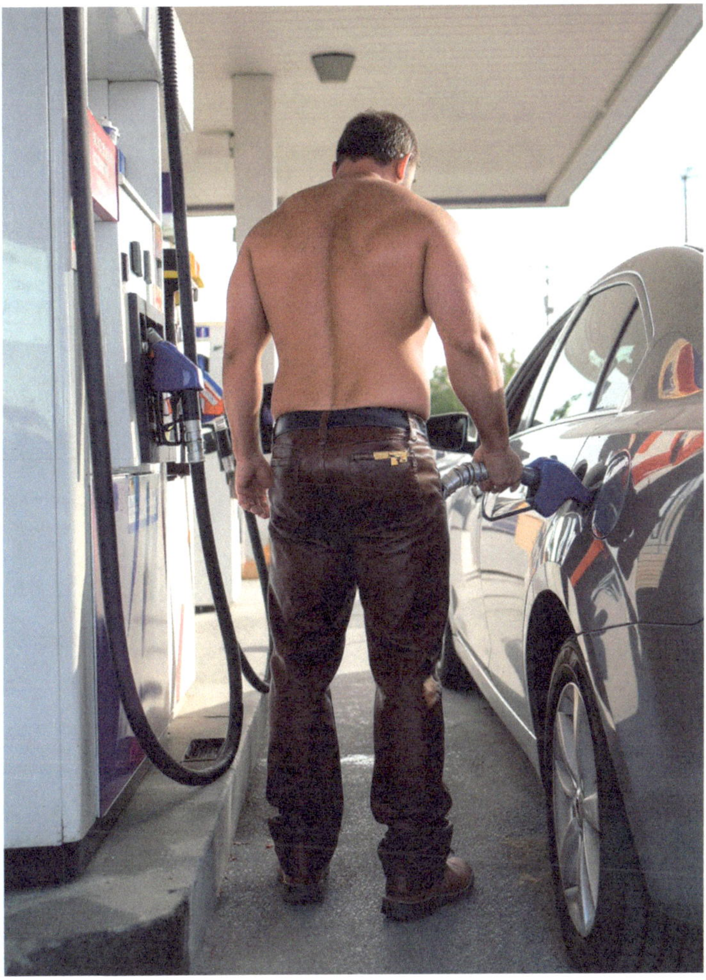

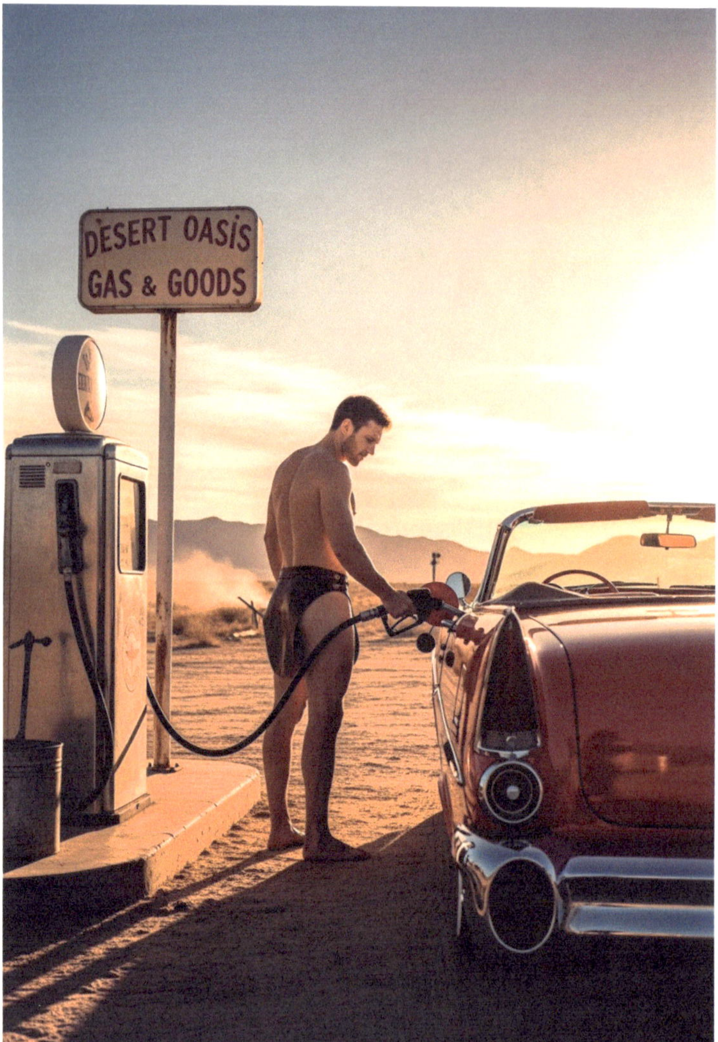
DESERT OASIS
GAS & GOODS

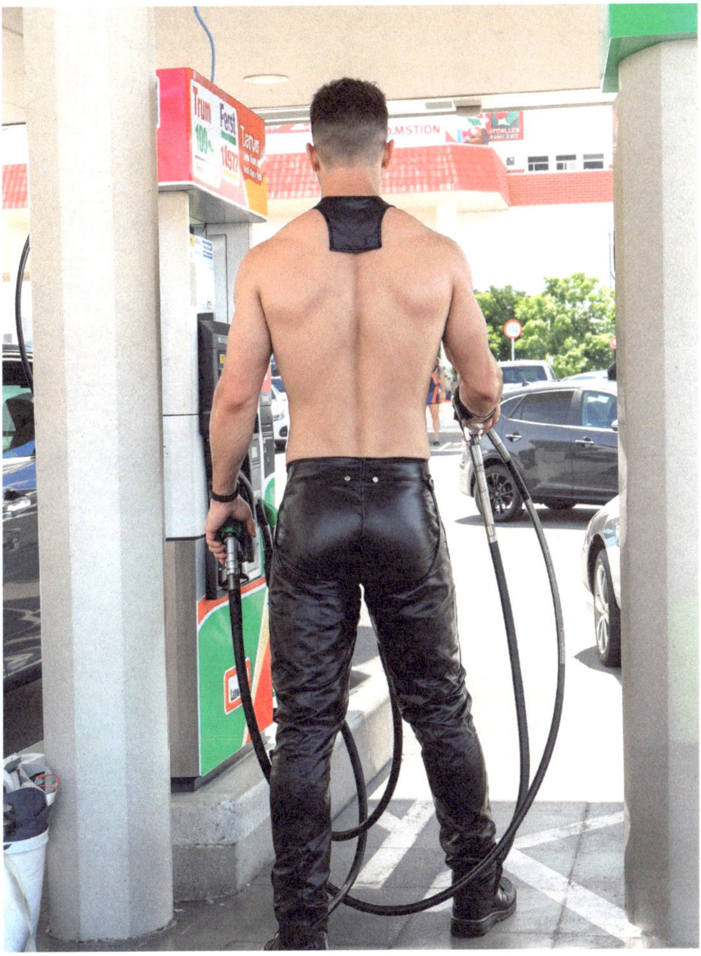

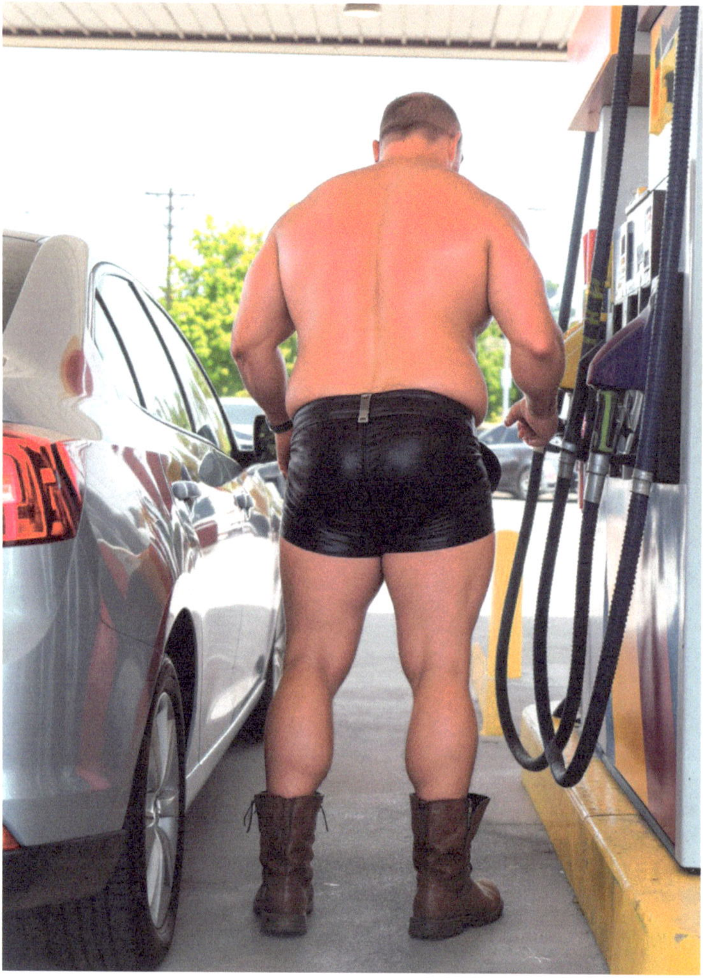

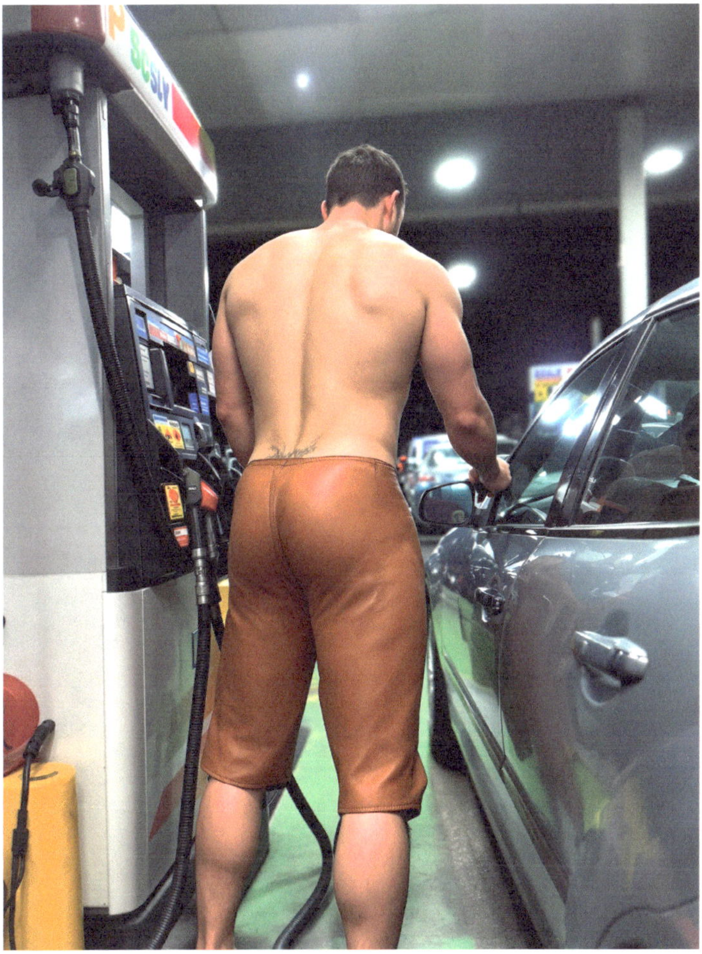

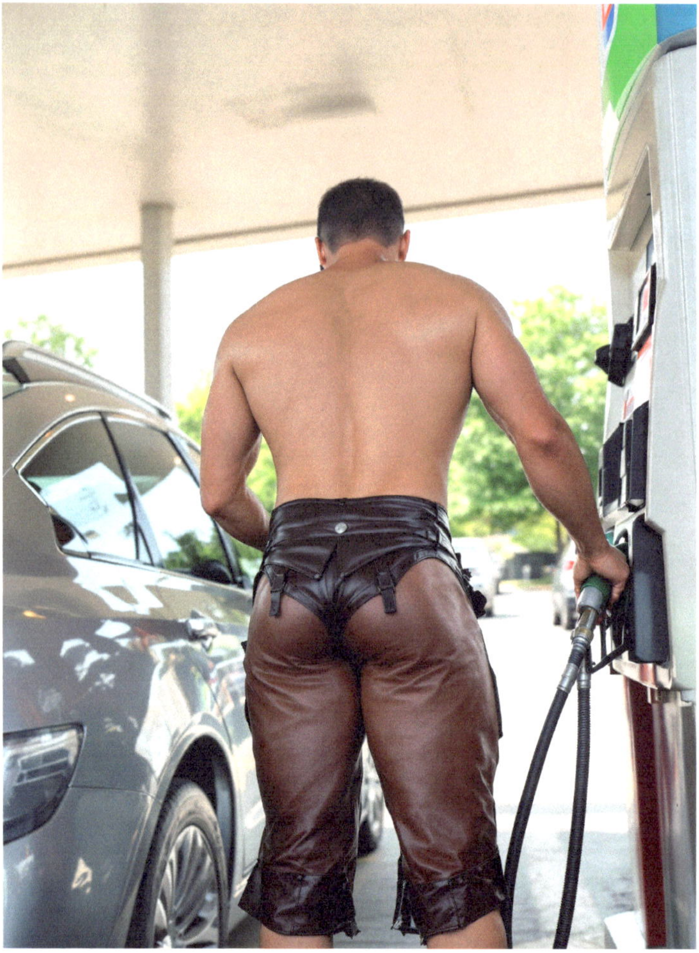

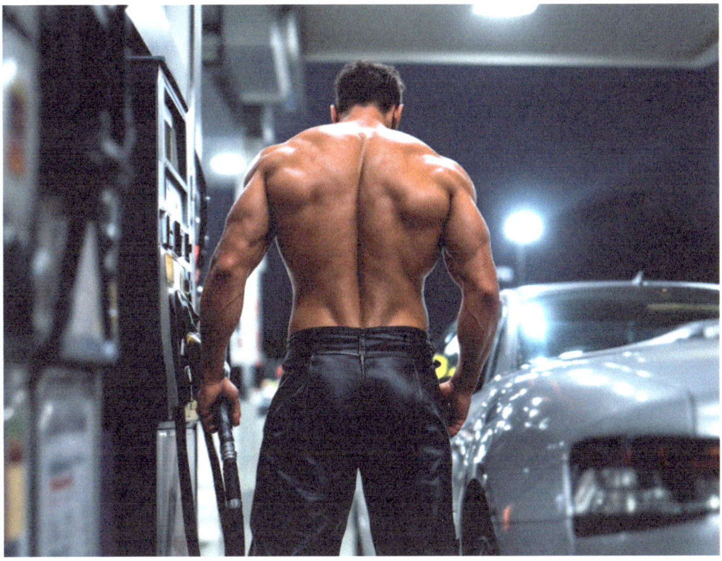

Text to image

full length shot of man in feather boa and high heels pumping gas at gas station

full length shot of man in feather boa a micro mini dress and high heels pumping gas at gas station

full length shot of man in feather boa a micro mini dress and high heels pumping gas at gas station

full length shot of man in feather boa a micro mini dress and high heels pumping gas at gas station

full length shot of man in feather boa a micro mini dress and high heels pumping gas at gas station

full length shot of man in feather boa a micro mini dress and high heels pumping gas at gas station full length shot of man in feather boa a micro mini dress and high heels pumping gas at gas station

full length shot of man in feather boa a micro mini dress and high heels pumping gas at gas station

full length shot of man in feather boa a micro mini dress and high heels pumping gas at gas station

full length shot of man in feather boa a micro mini dress
and high heels pumping gas at gas station

full length shot of man in feather boa a micro mini dress
and high heels pumping gas at gas station

full length shot of man in feather boa a micro mini dress
and high heels pumping gas at gas station

full length shot from above of man in feather boa a micro
mini dress and high heels pumping gas at gas station

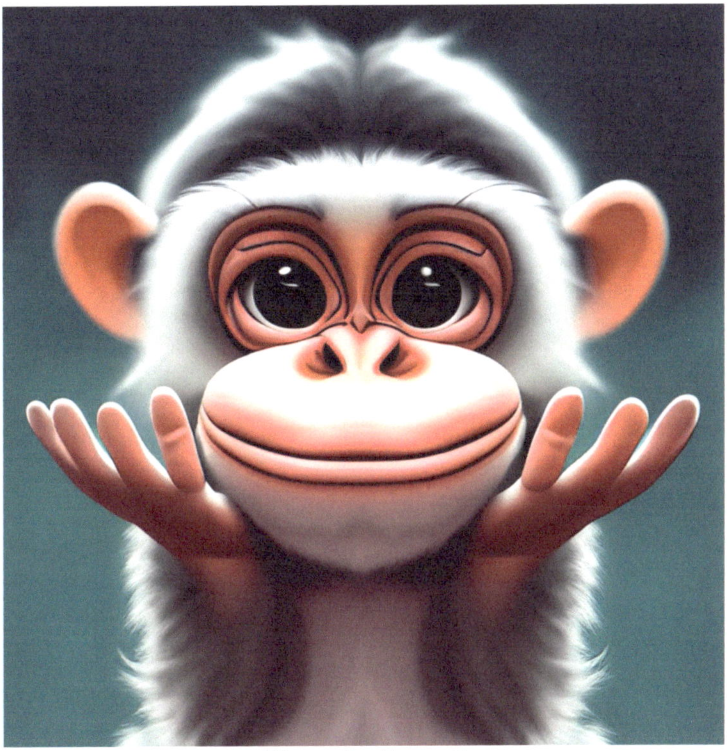

Can't load

We can't display the generated image.
You can edit and try again.

full length shot from above of man in feather boa a micro mini dress and high heels pumping gas at gas station

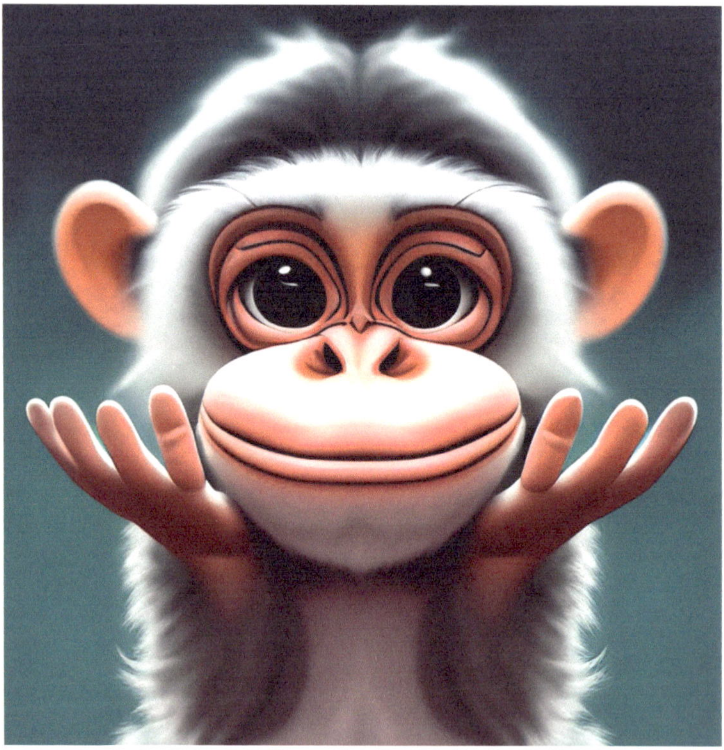

Can't load

We can't display the generated image.
You can edit and try again.

full length shot from above of man in feather boa a micro mini dress and high heels pumping gas at gas station

full length shot from above of man in feather boa a micro mini dress and high heels pumping gas at gas station

full length shot of man in feather boa a micro mini dress and high heels pumping gas at gas station

full length shot of man in feather boa a micro mini dress and high heels pumping gas at gas station

full length shot from below of man in feather boa a micro mini dress and high heels pumping gas at gas station

full length shot from below of man in feather boa a micro mini dress and high heels pumping gas at gas station

full length shot from below of man in feather boa a micro mini dress and high heels pumping gas at gas station

full length shot from below of man in feather boa a micro mini dress and high heels pumping gas at gas station

full length shot from below of man in feather boa a micro mini dress and high heels pumping gas at gas station

full length shot from below of man in feather boa a micro mini dress and high heels pumping gas at gas station

full length shot from below of man in feather boa a micro mini dress and high heels pumping gas at gas station

full length shot from below of man in feather boa a micro mini dress and high heels pumping gas at gas station

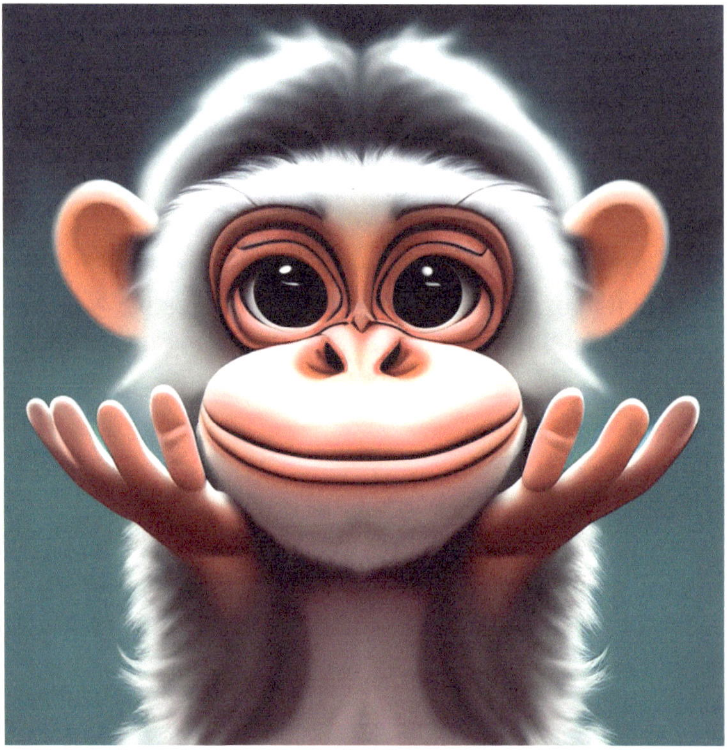

Can't load

We can't display the generated image.
You can edit and try again.

full length shot from below of man in feather boa a micro mini dress and high heels pumping gas at gas station

full length shot from below of man in feather boa a micro mini dress and high heels pumping gas at gas station

full length shot from below of man in feather boa a micro mini dress and high heels pumping gas at gas station

full length shot from below of man in feather boa a micro mini dress and high heels pumping gas at gas station

full length shot from below of man in feather boa a micro mini dress and high heels pumping gas at gas station

full length shot from below of man in feather boa a micro mini dress and high heels pumping gas at gas station

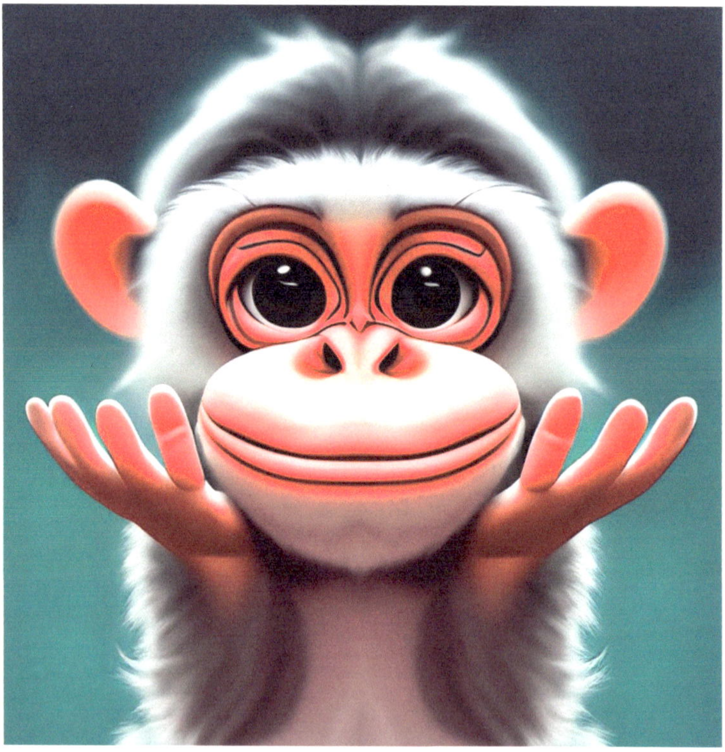

Can't load

We can't display the generated image.
You can edit and try again.

gentleman pumping gas into vehicle at gas station

gentleman pumping gas into vehicle at gas station

gentleman pumping gas into vehicle at gas station

gentleman pumping gas into vehicle at gas station

gentleman pumping gas into vehicle at gas station

gentleman pumping gas into vehicle at gas station

gentleman pumping gas into vehicle at gas station

man with bird on his shoulder pumping gas at gas station

man with bird on his shoulder pumping gas at gas station

man with bird on his shoulder pumping gas at gas station

man with bird on his shoulder pumping gas at gas station

man with bird on his shoulder pumping gas at gas station

man with bird on his shoulder pumping gas at gas station

man with bird on his shoulder pumping gas at gas station

studly man pumping gas at a gas station

studly man pumping gas at a gas station

studly man pumping gas at a gas station

f

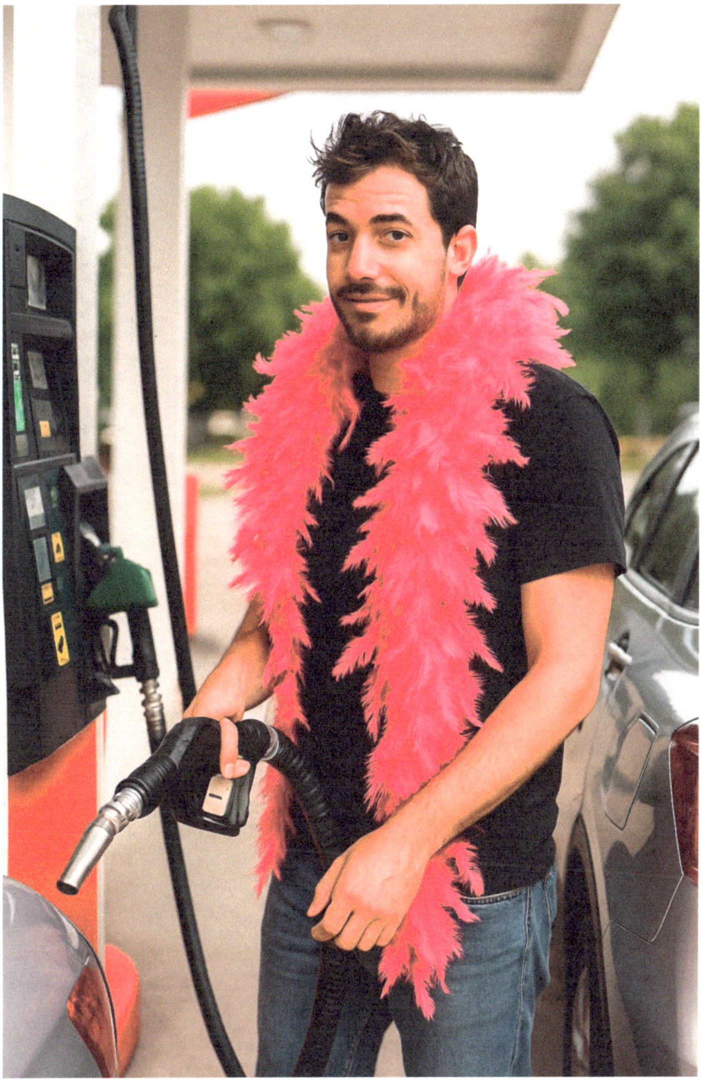

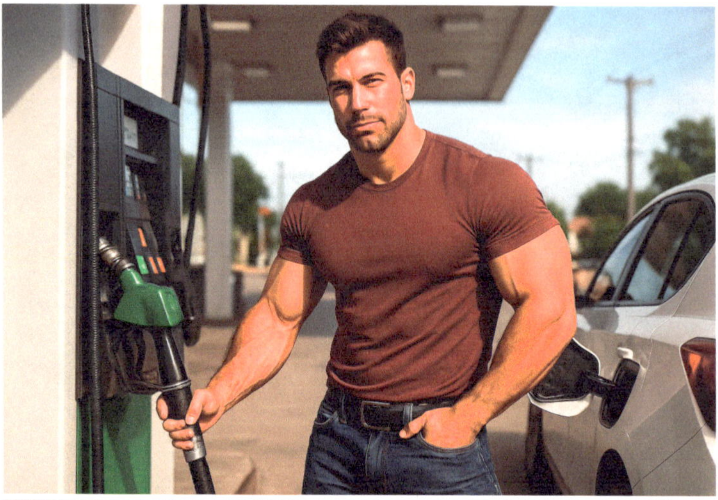

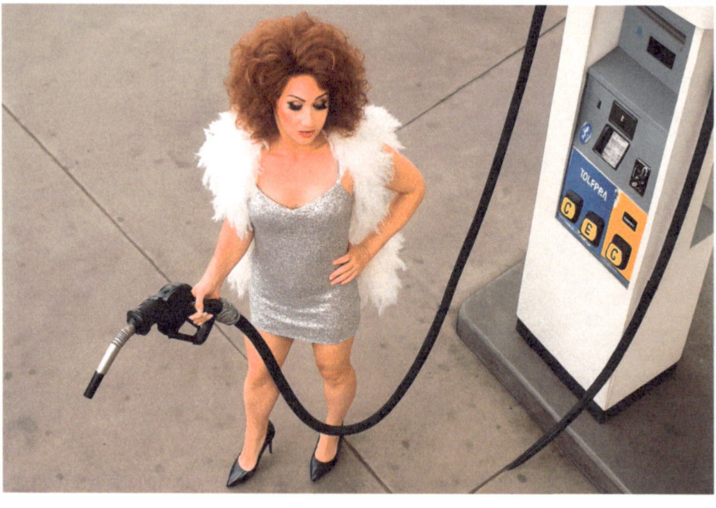

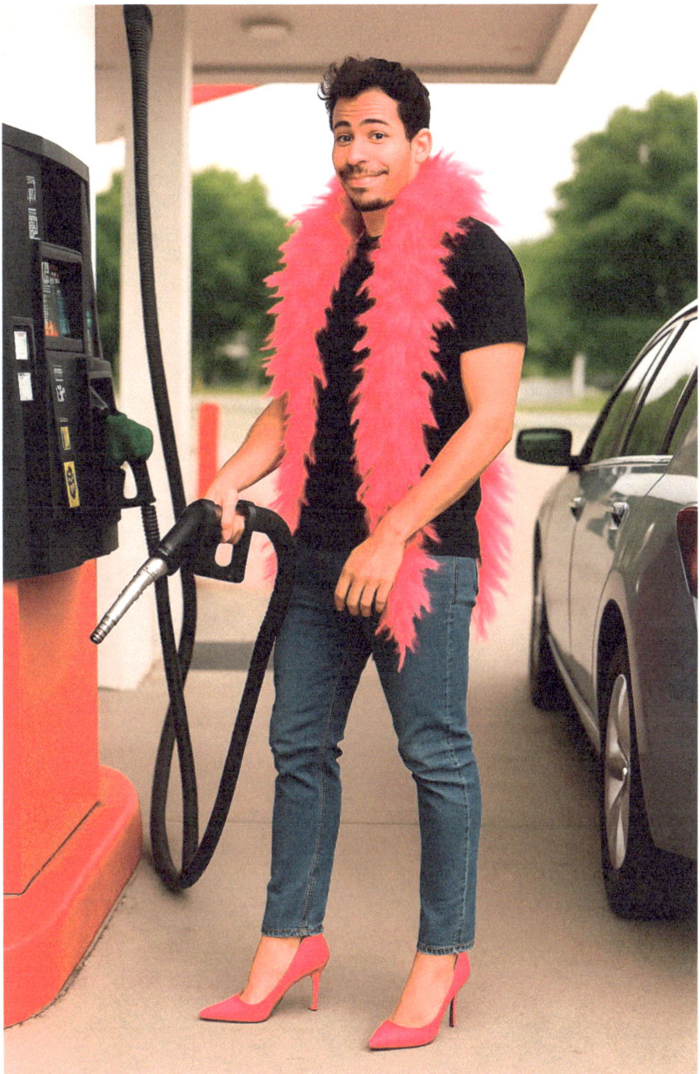

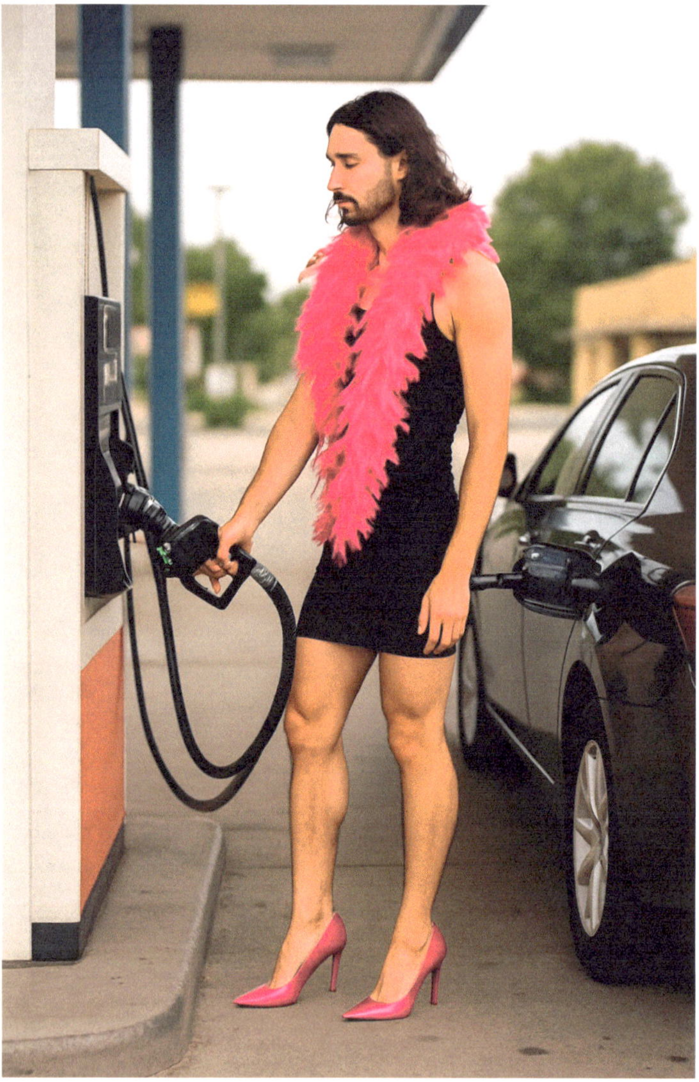

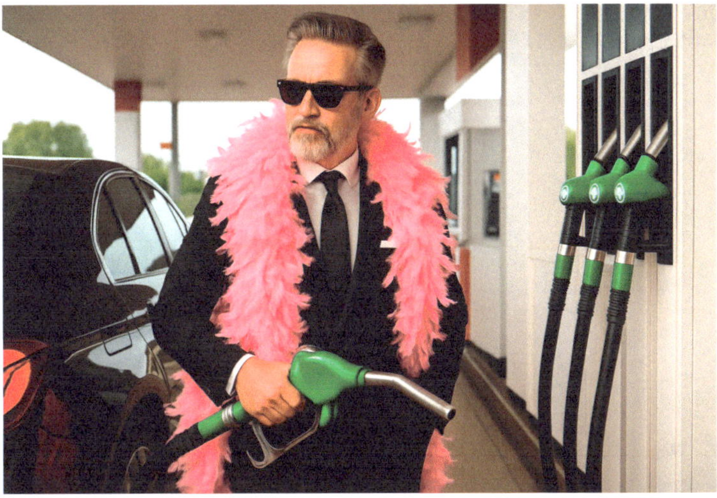

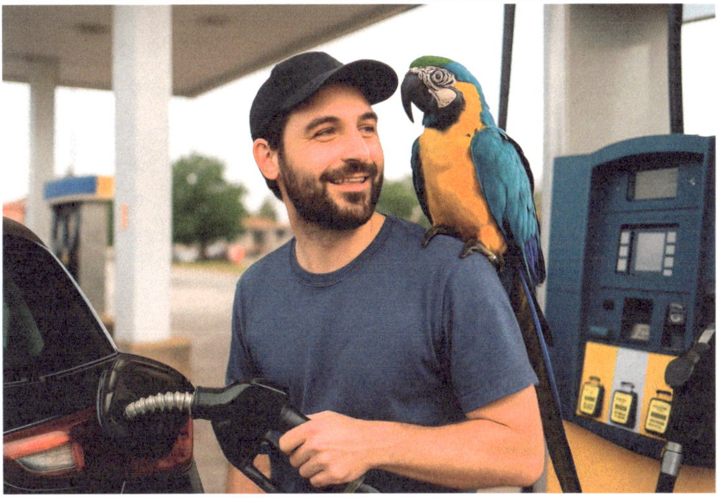

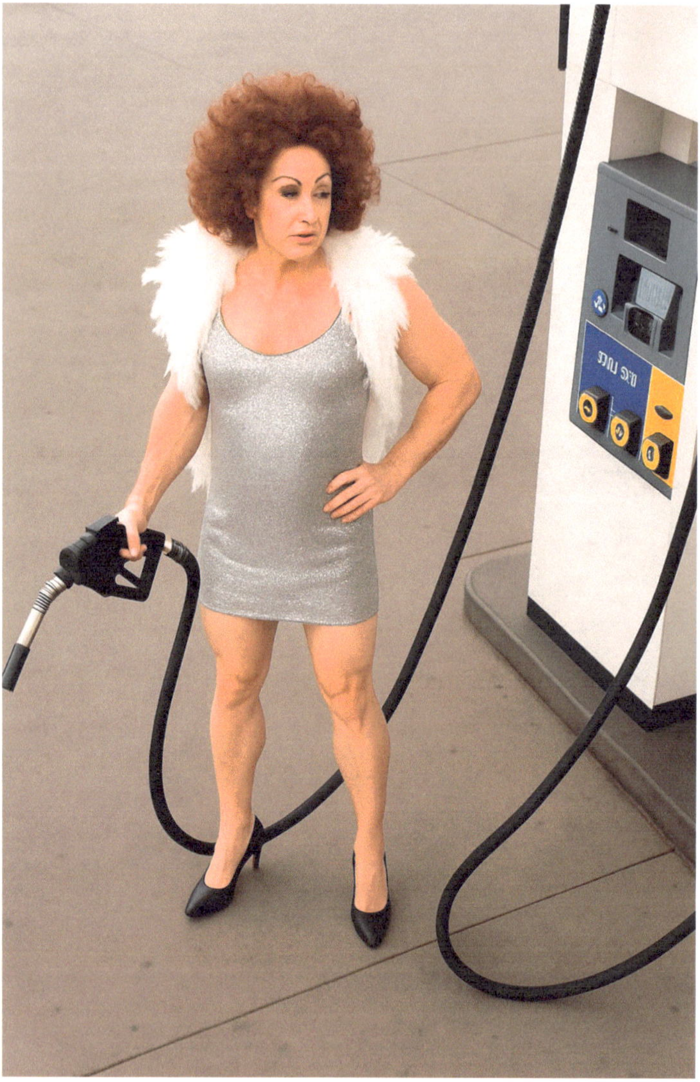

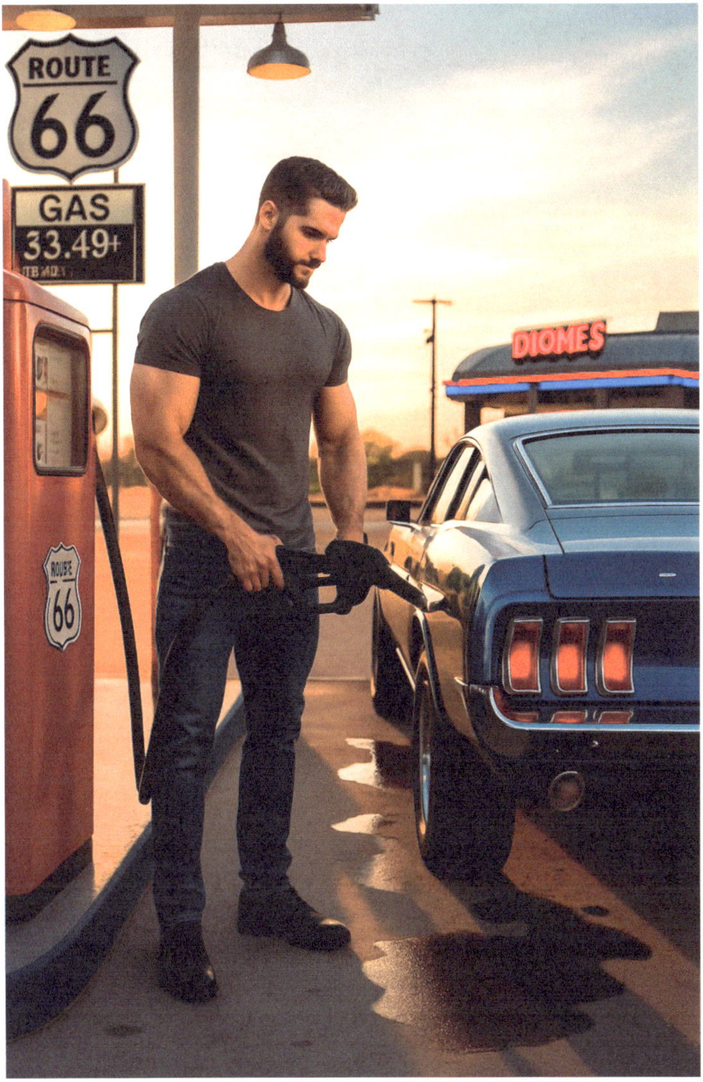
ROUTE
66
GAS
33.49+
DIOMES
66

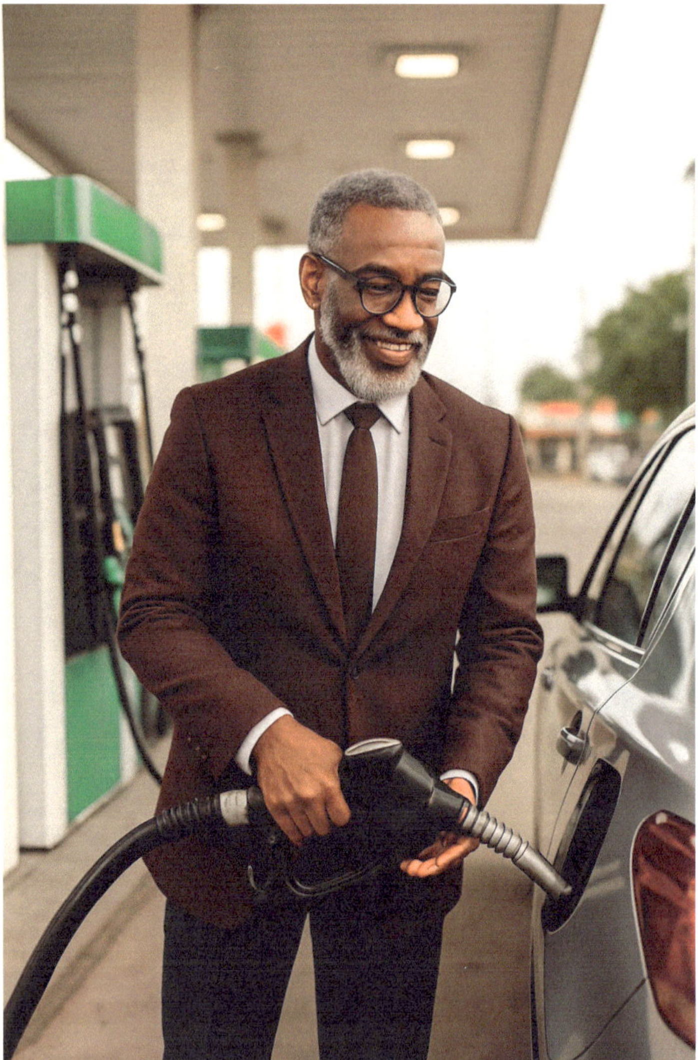

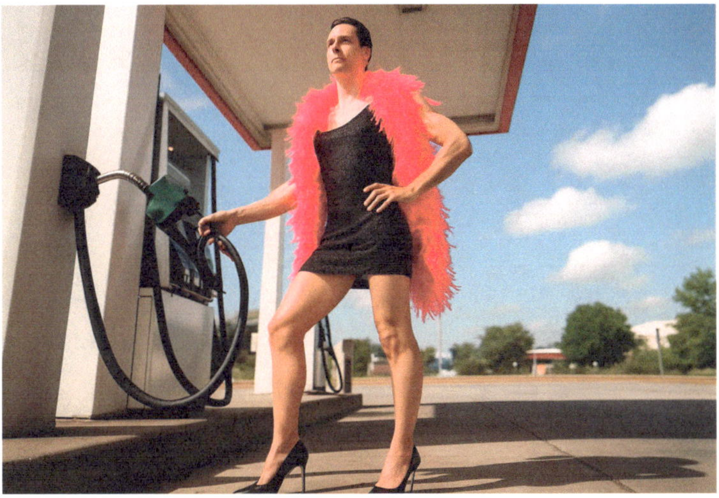

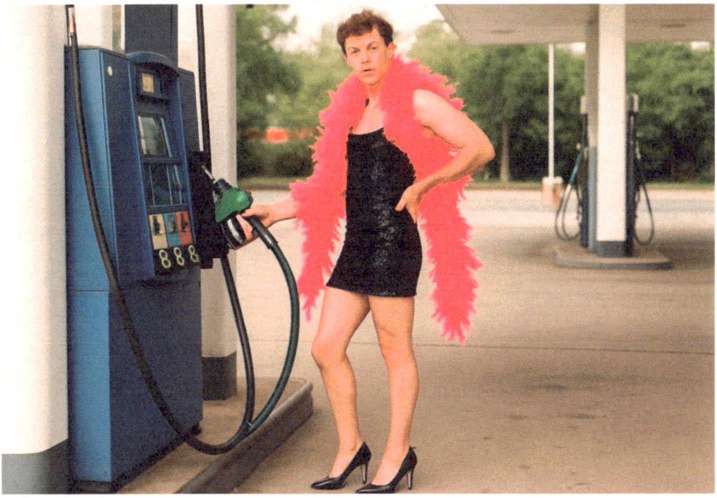

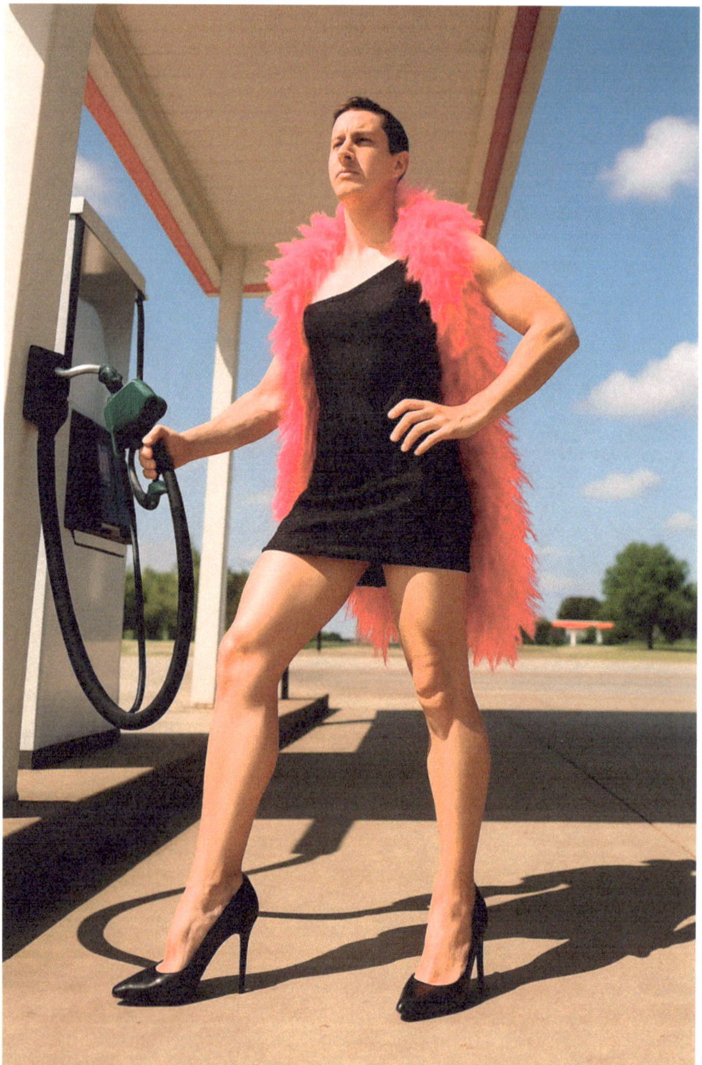

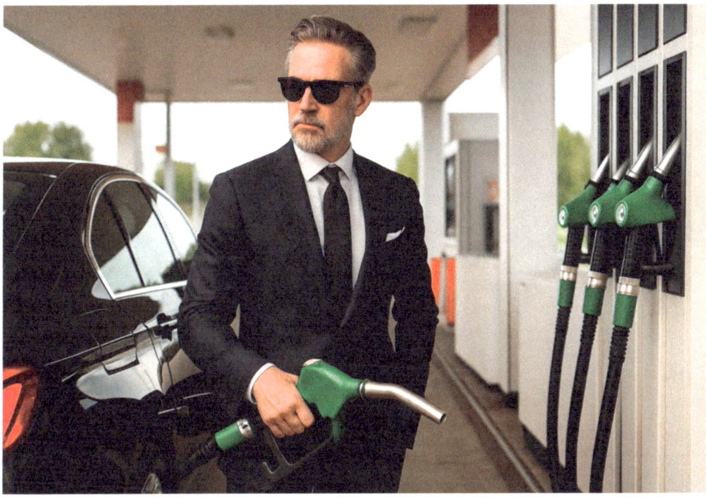

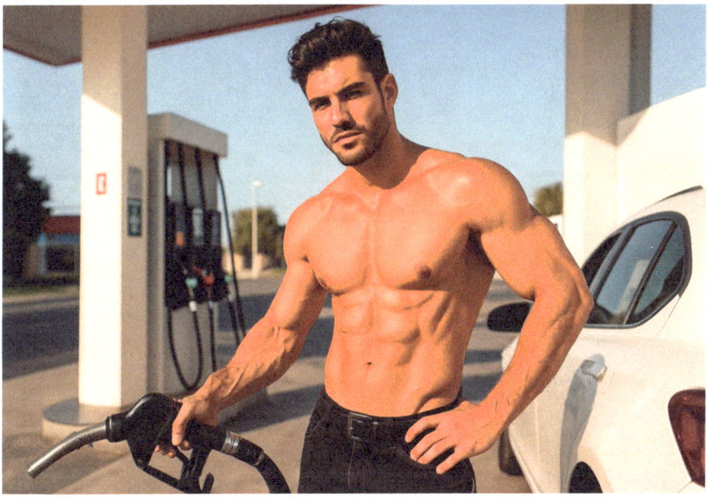

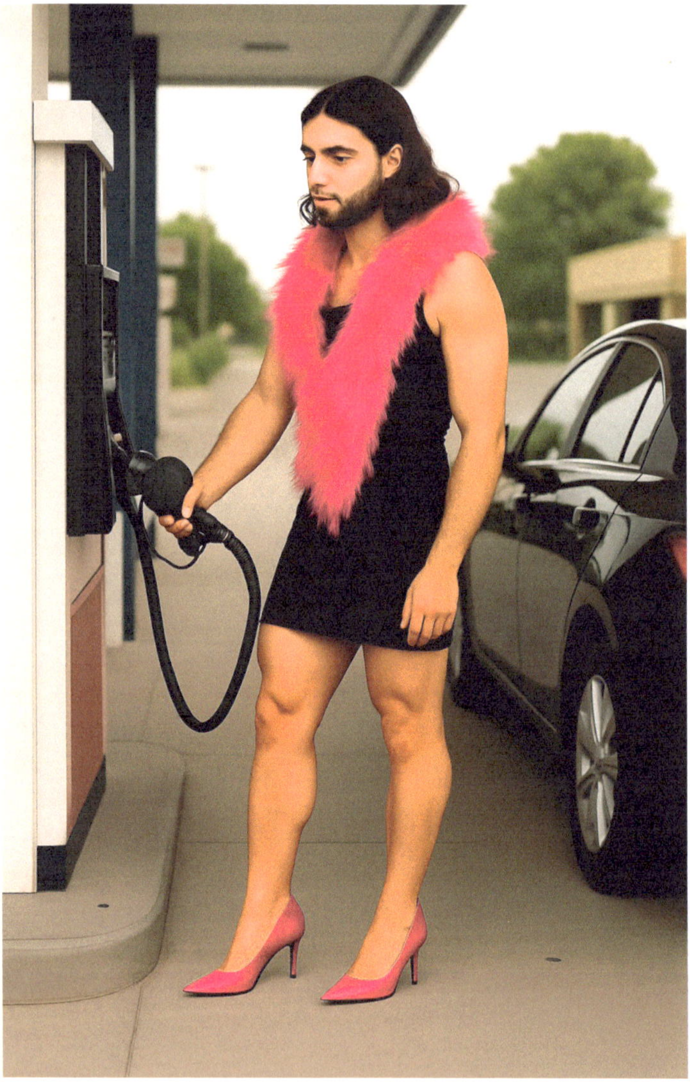

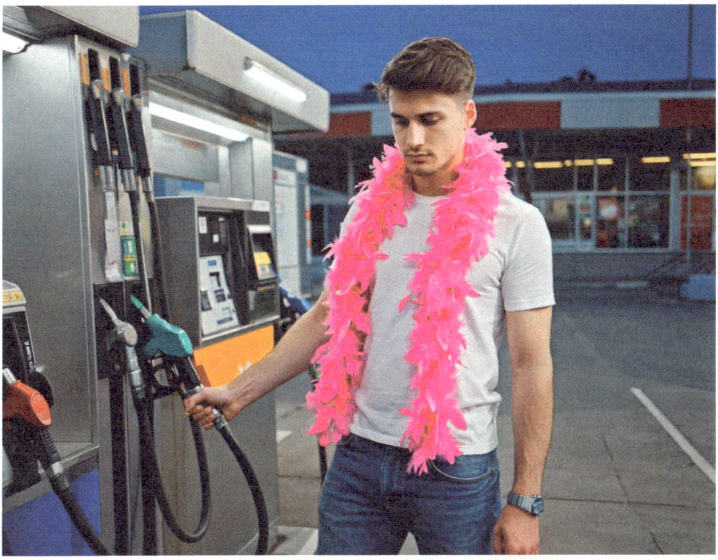

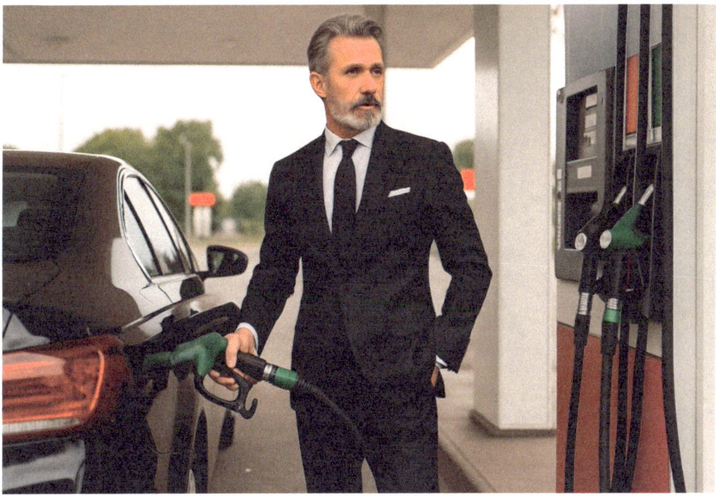

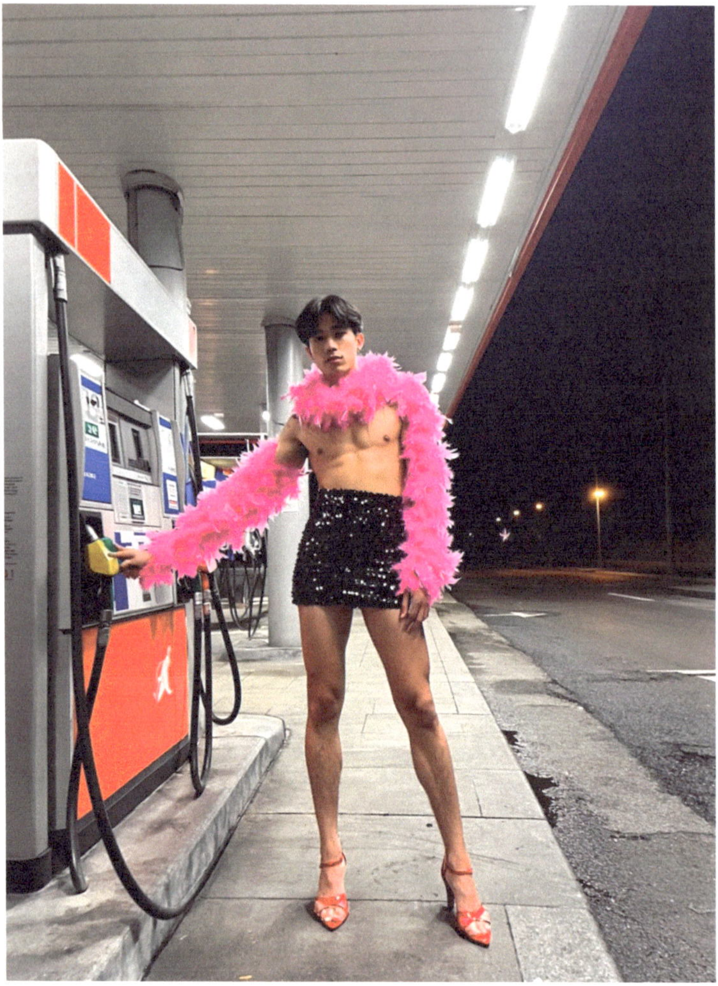

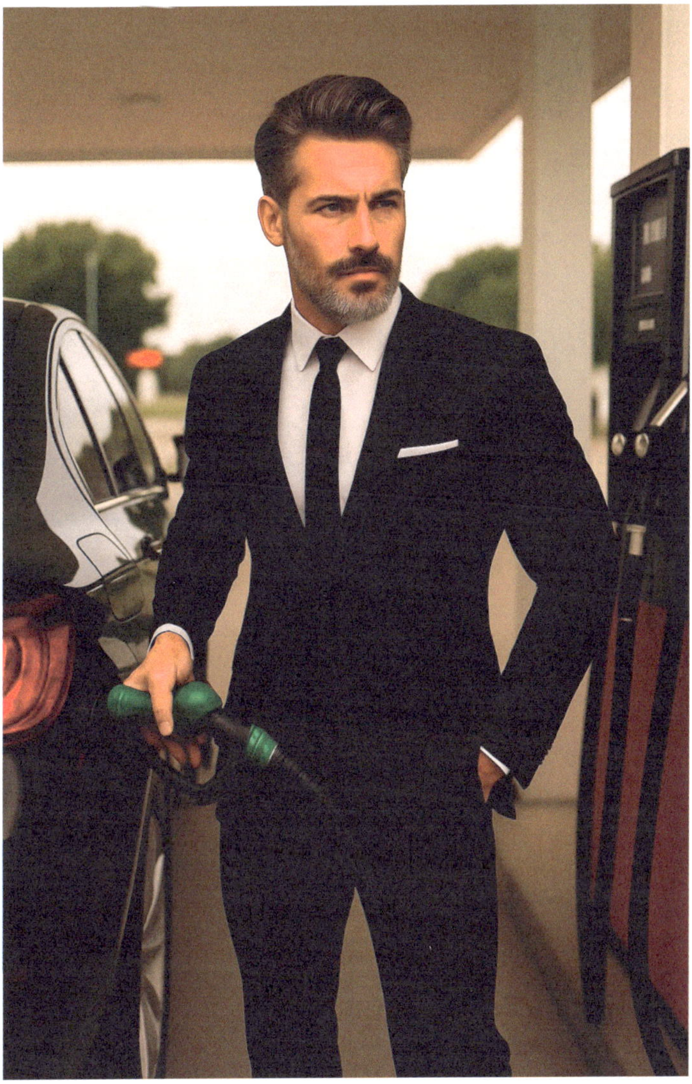

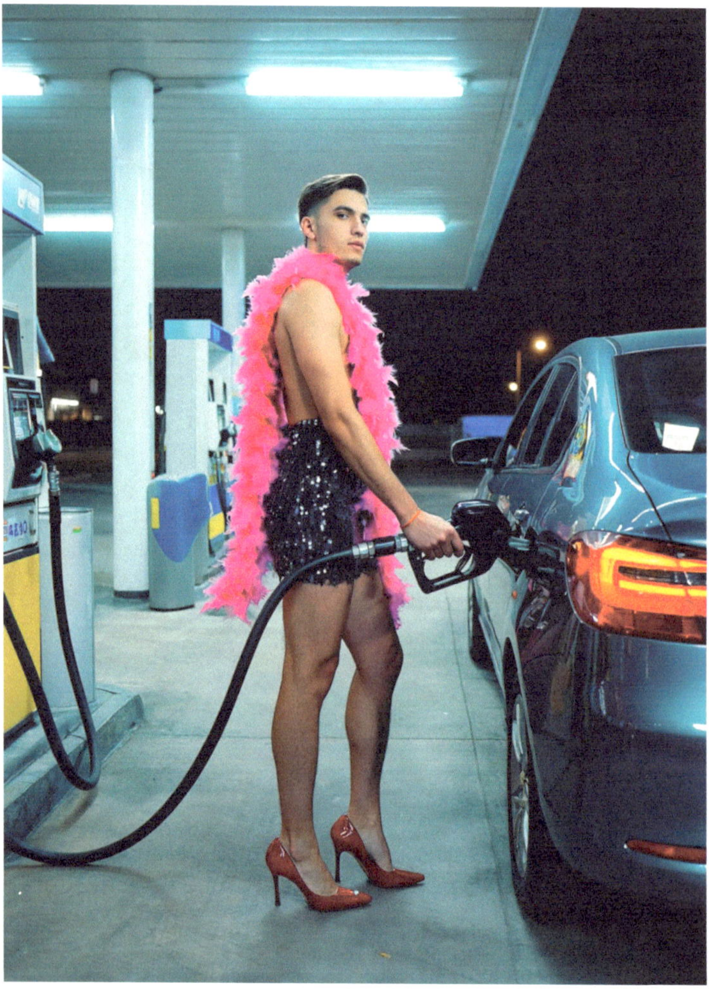

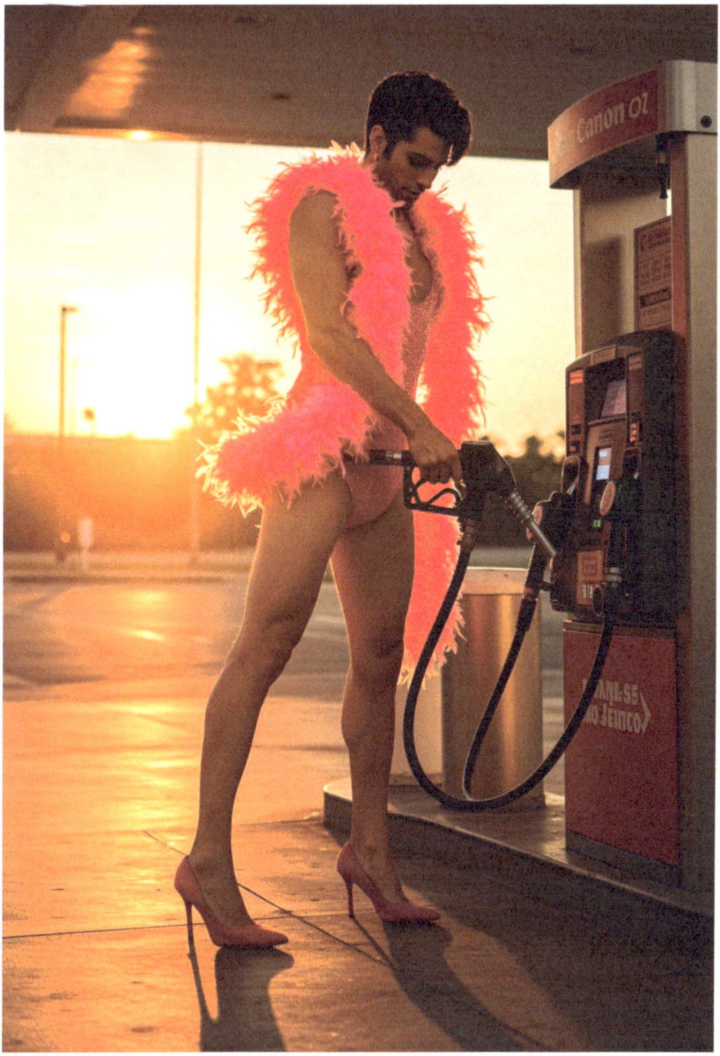
Canon

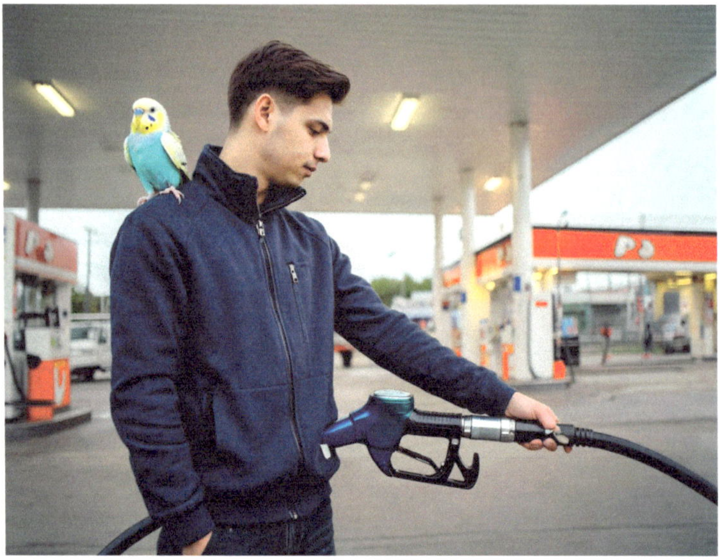

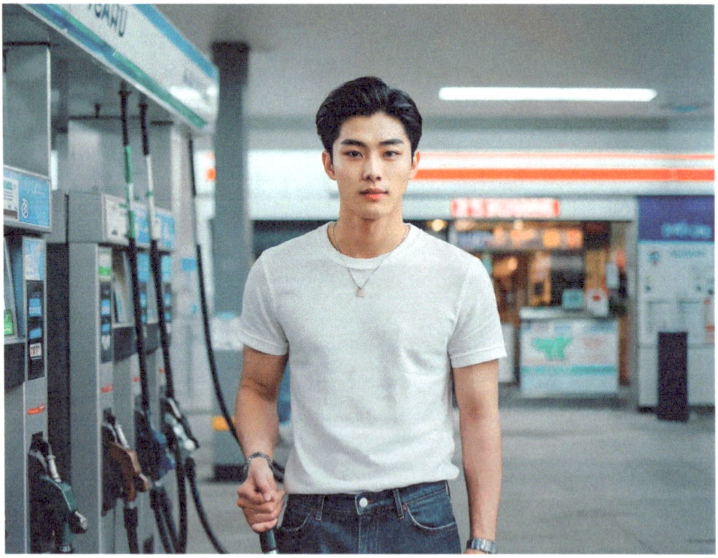

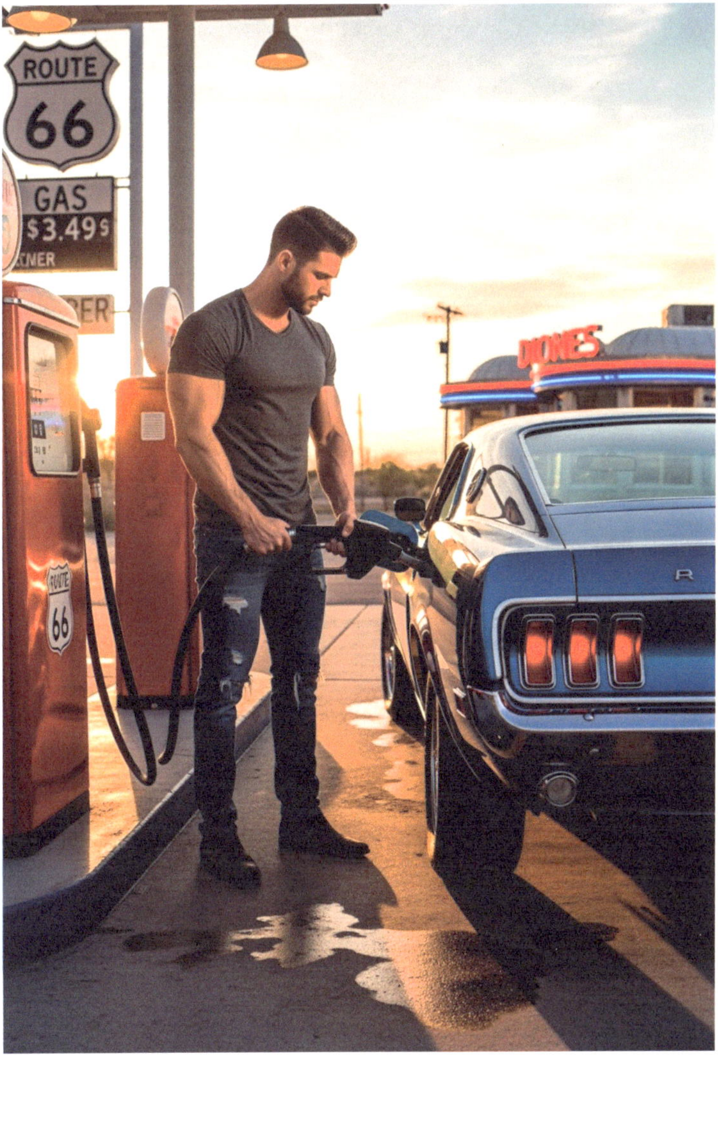
ROUTE
66
GAS
$3.49
ROUTE
66

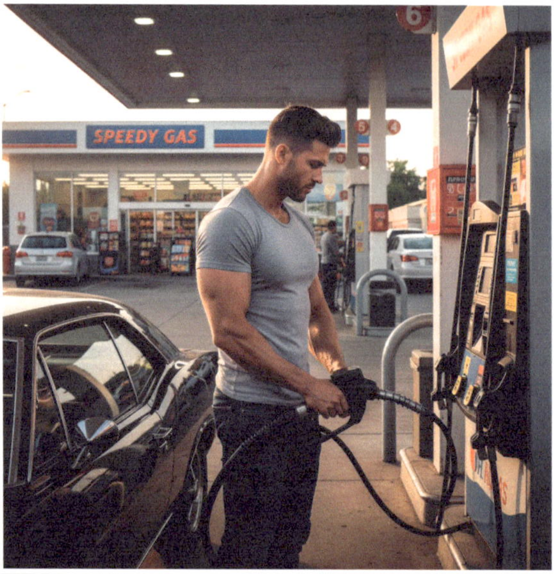
SPEEDY GAS

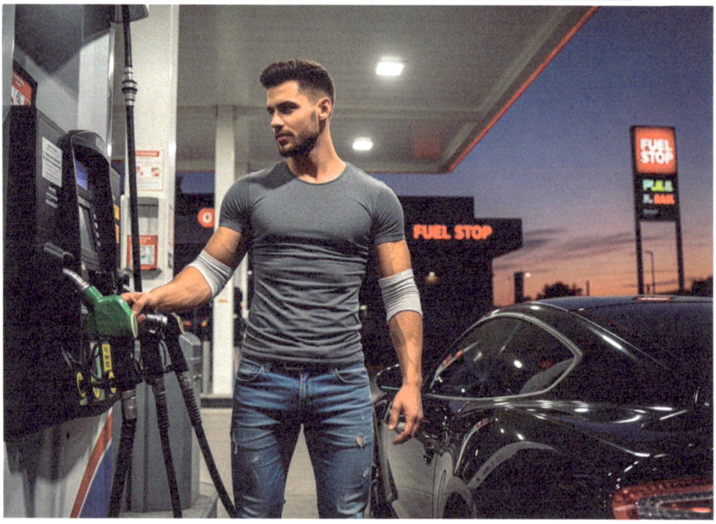
FUEL STOP
FUEL STOP

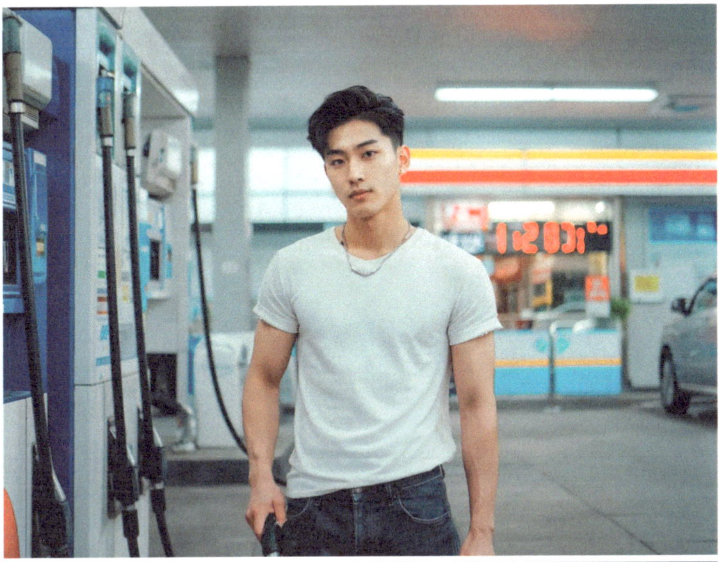

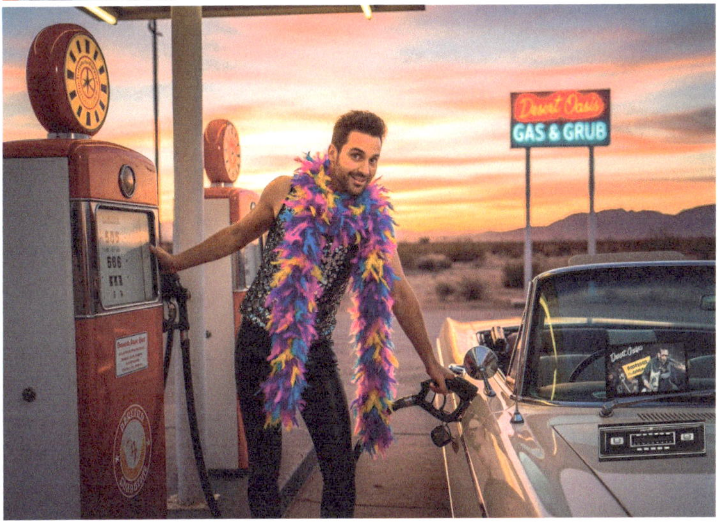
Desert Oasis
GAS & GRUB

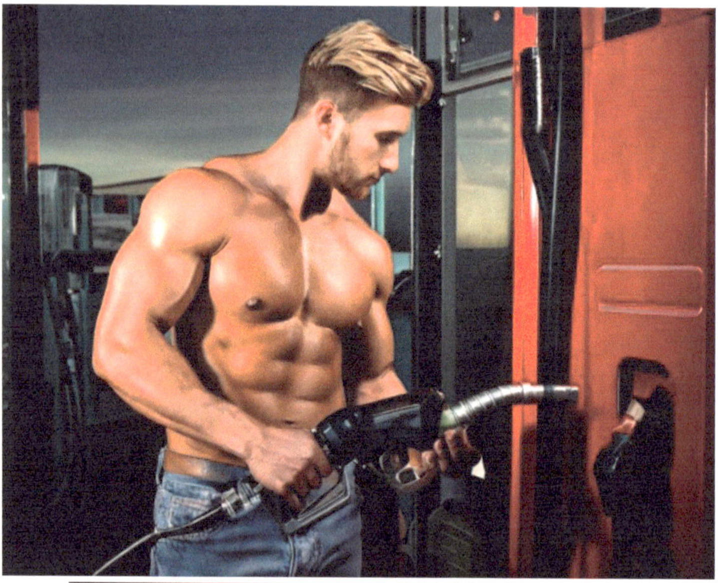

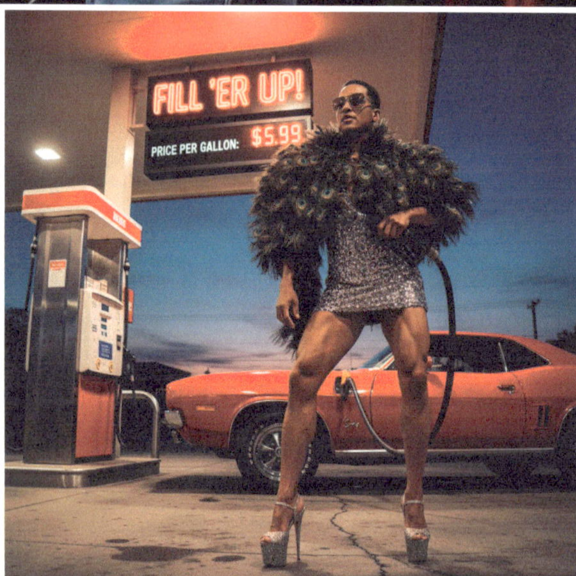
FILL 'ER UP!
PRICE PER GALLON: $5.99

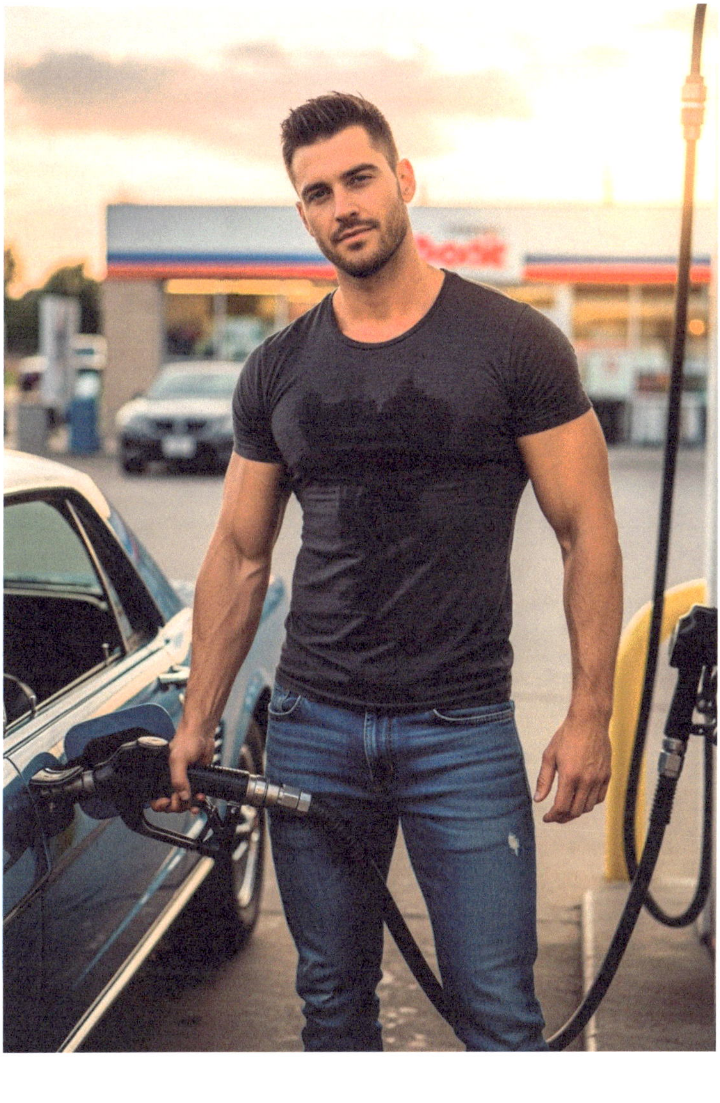

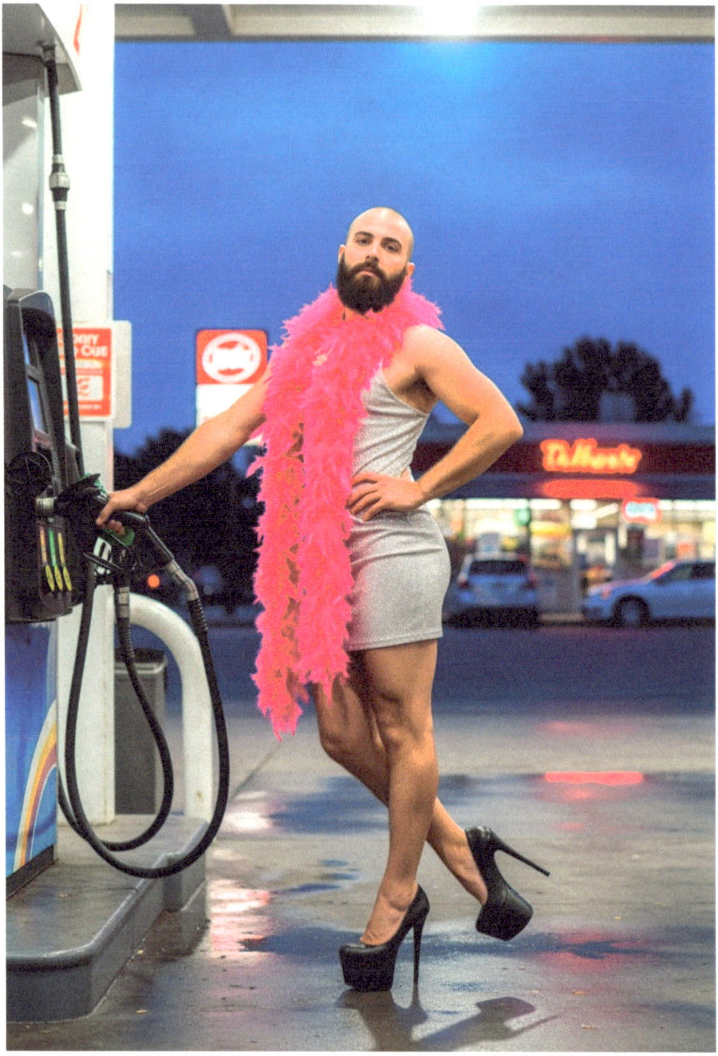

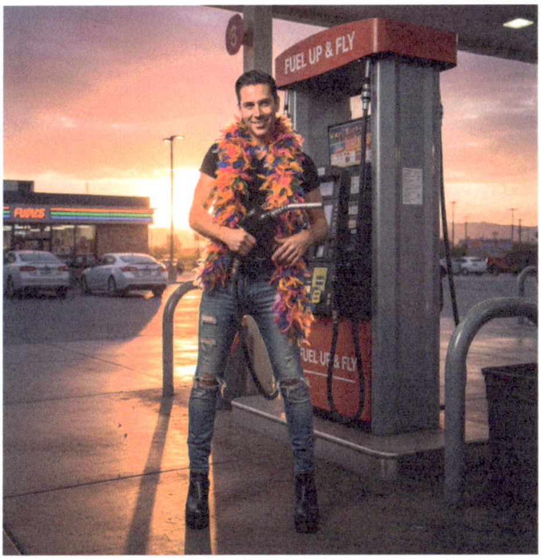
FUEL UP & FLY
FUEL UP & FLY

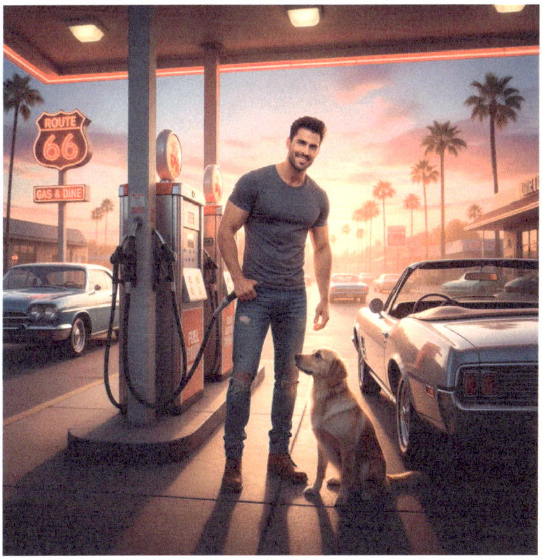
ROUTE
66
GAS & DINE
FUEL

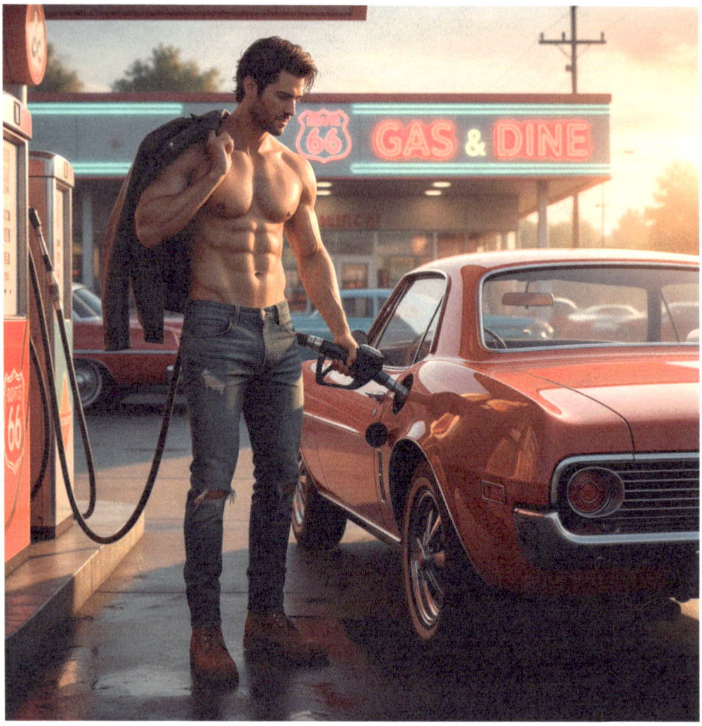
66
GAS & DINE

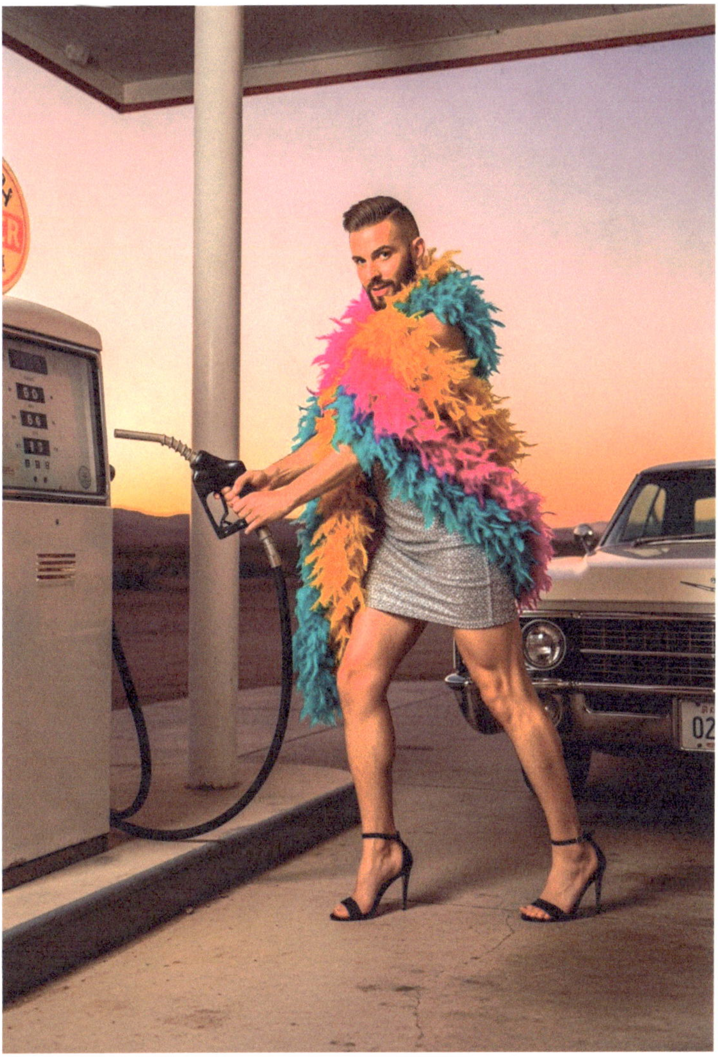

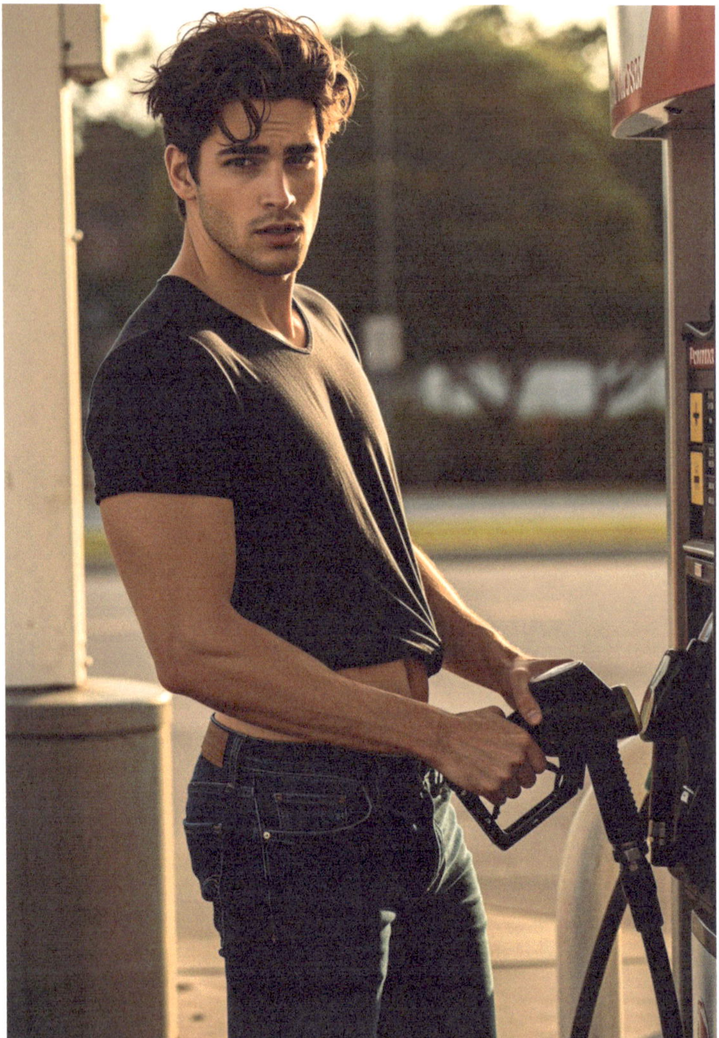

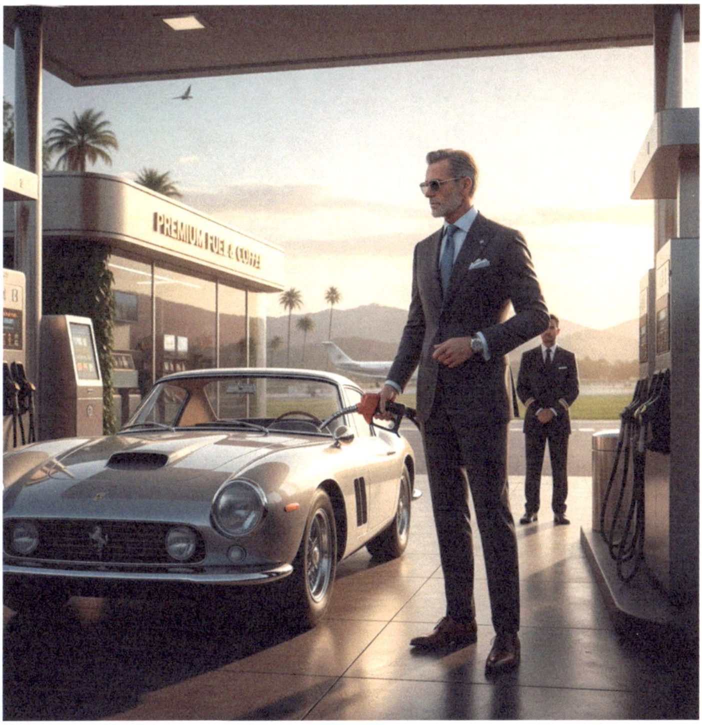
PREMIUM

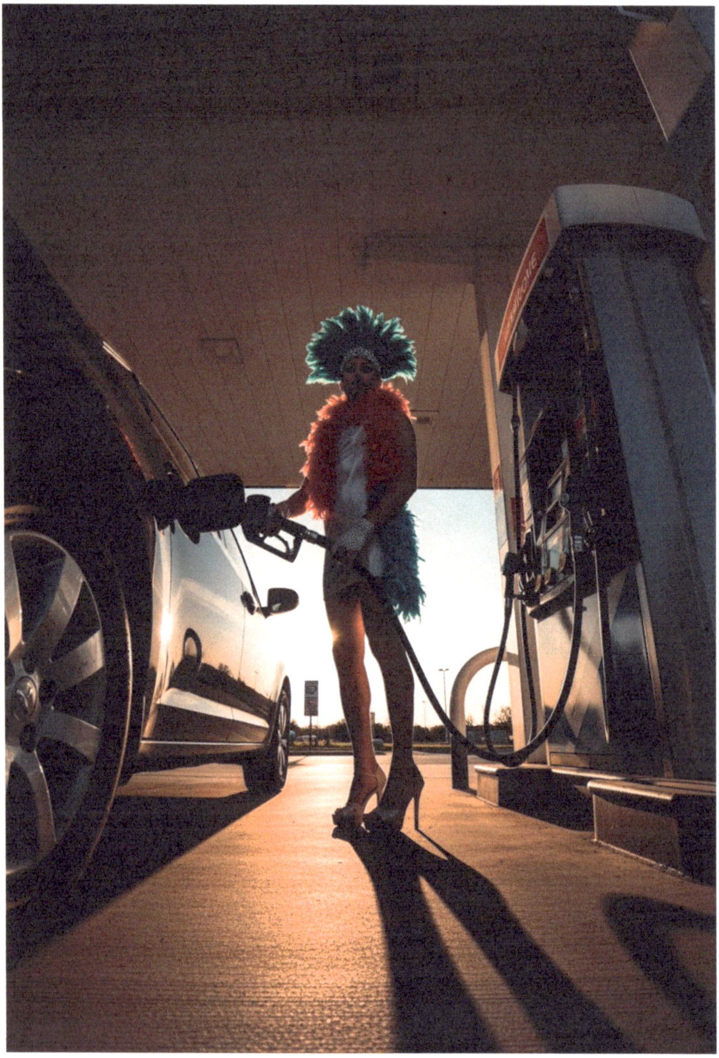

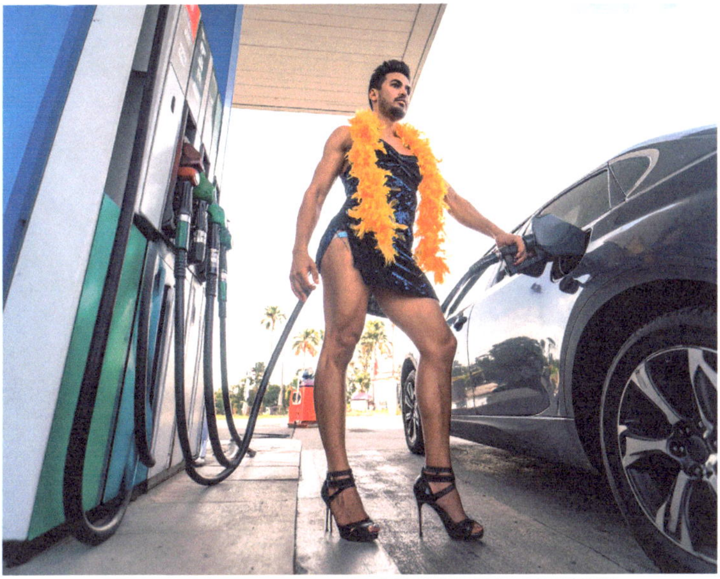

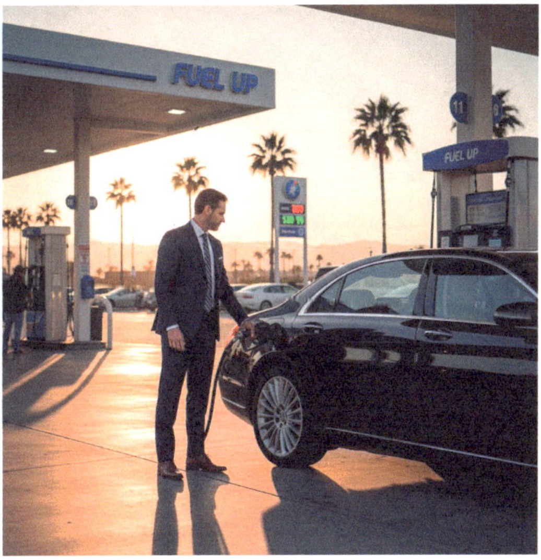
FUEL UP
FUEL UP

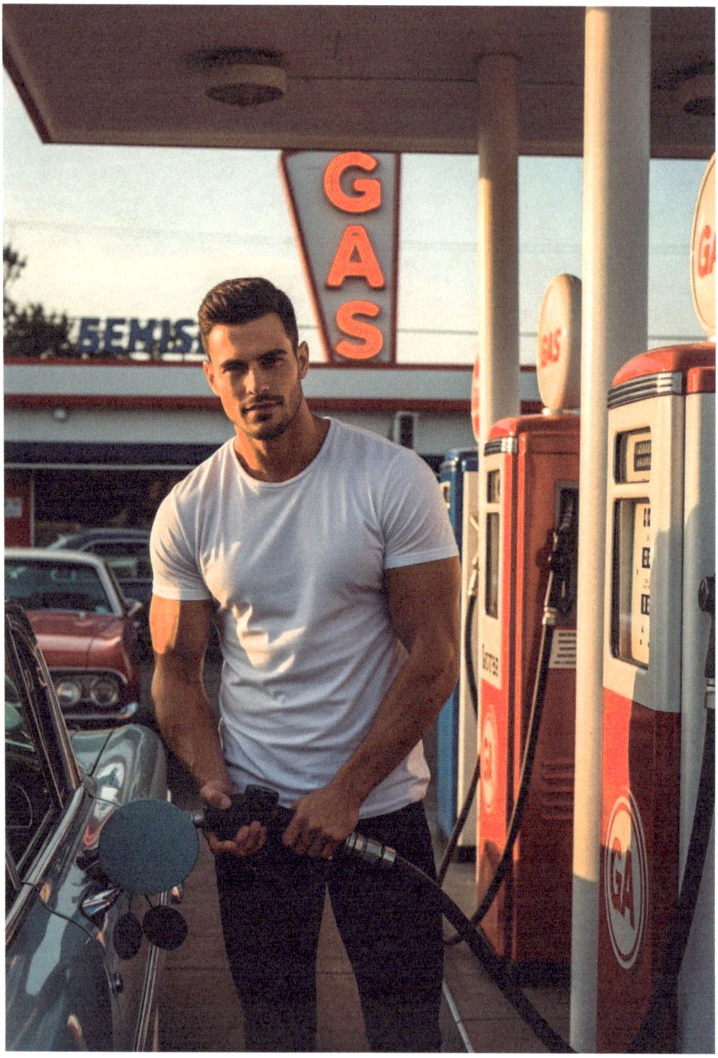
GAS

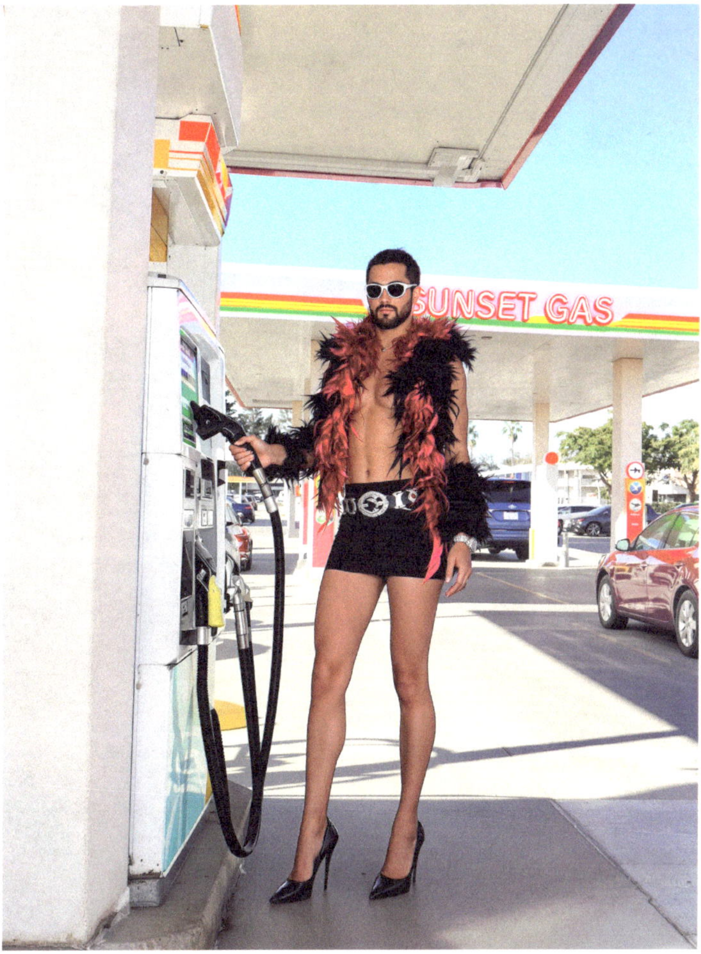
SUNSET GAS

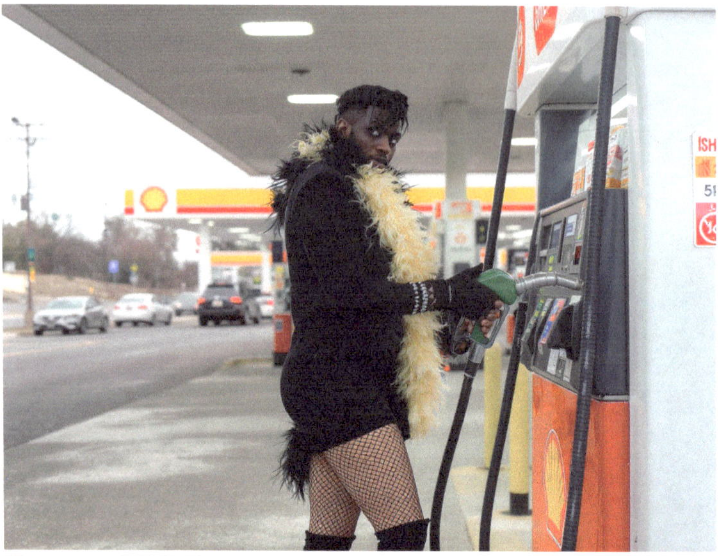

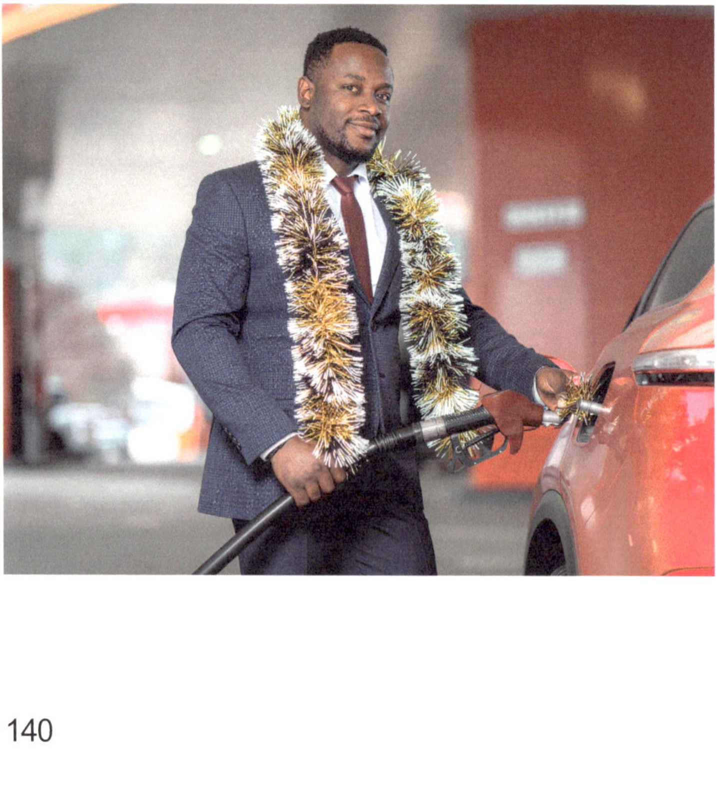

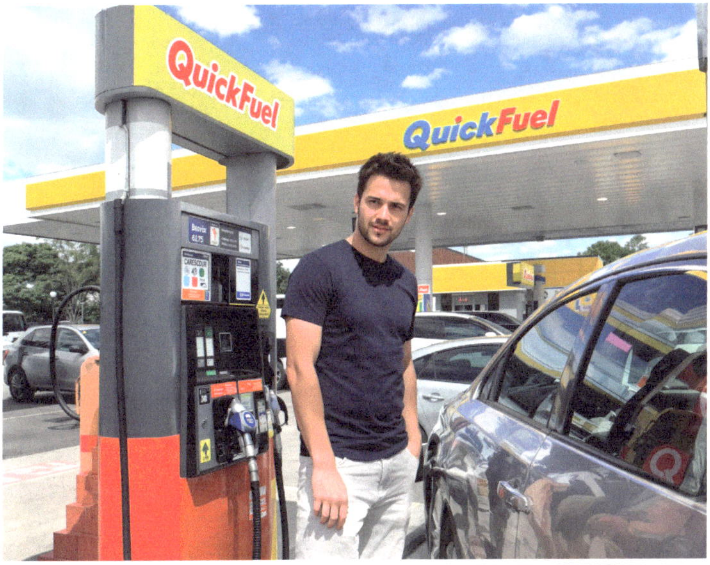
QuickFuel
QuickFuel

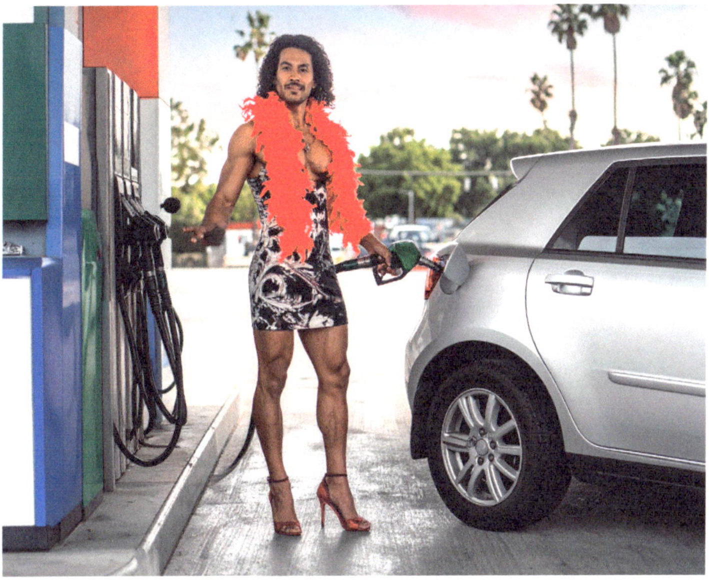

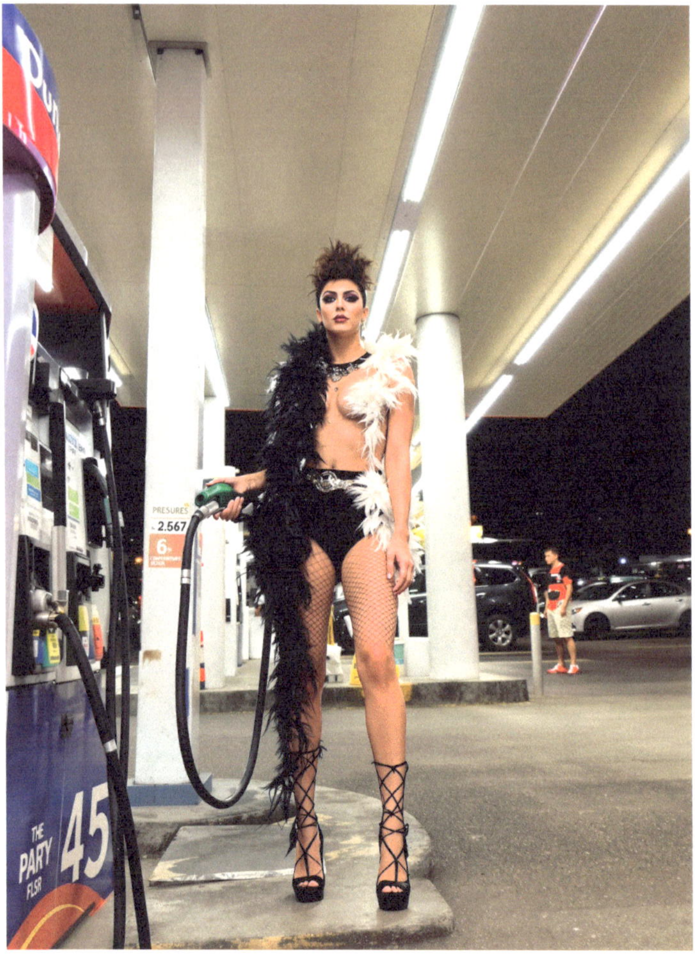
2.567
THE
PARTY
45

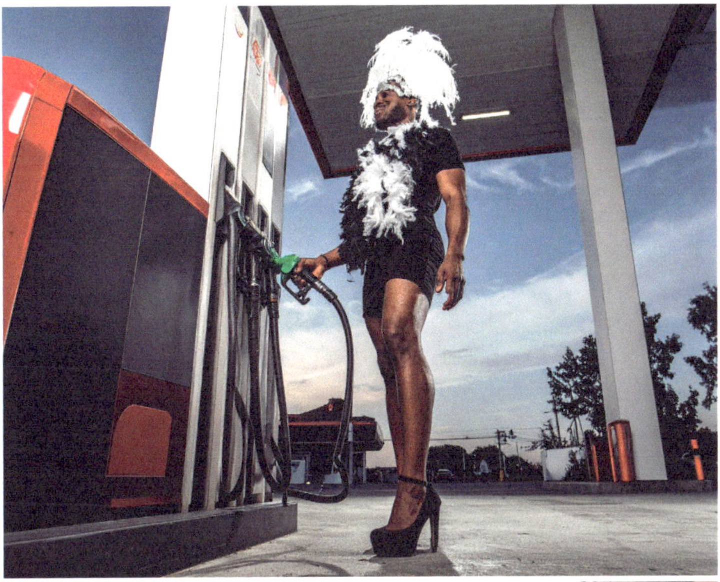

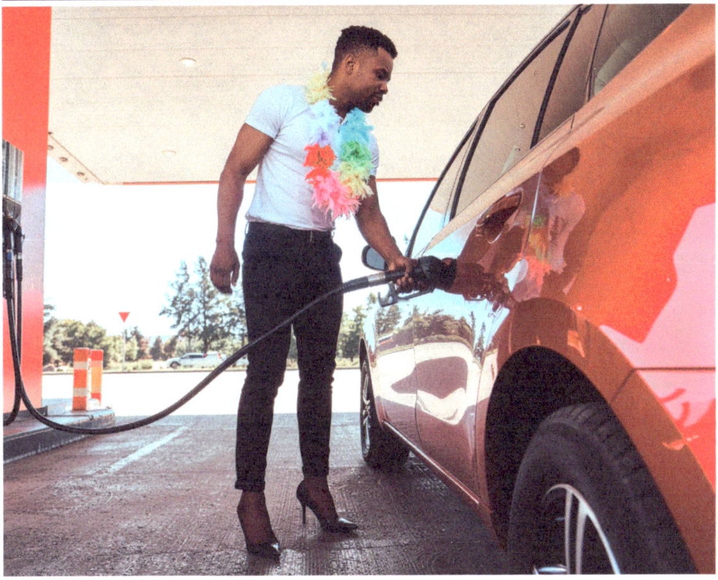

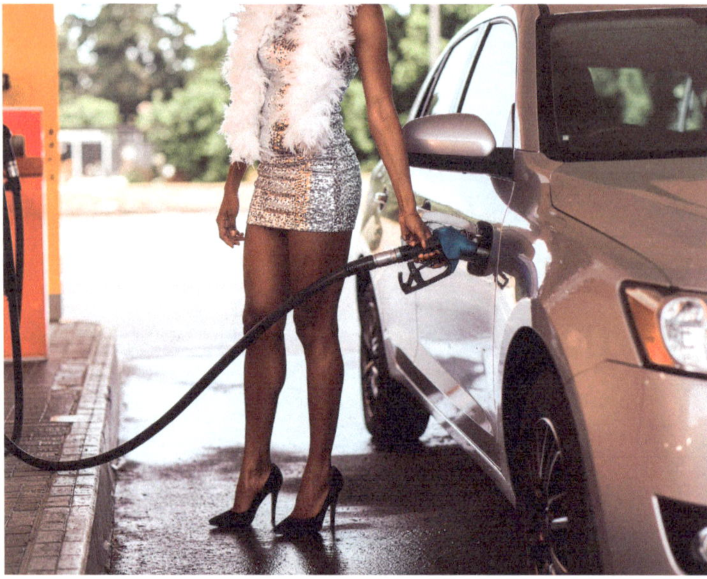

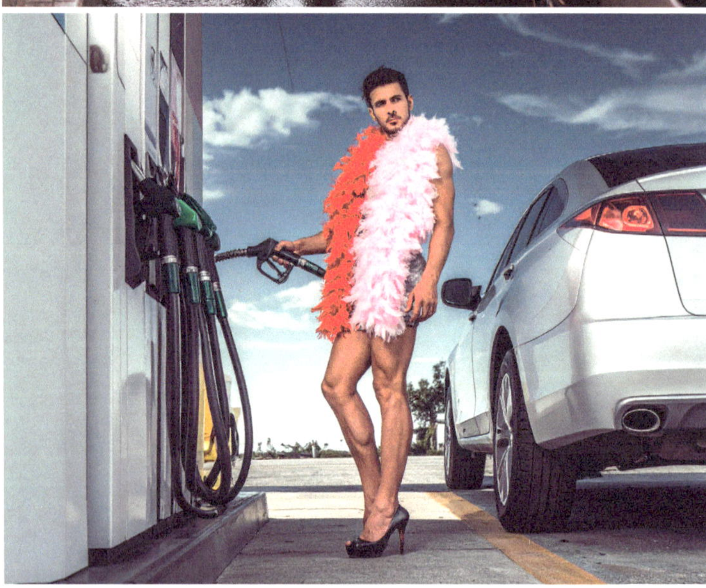

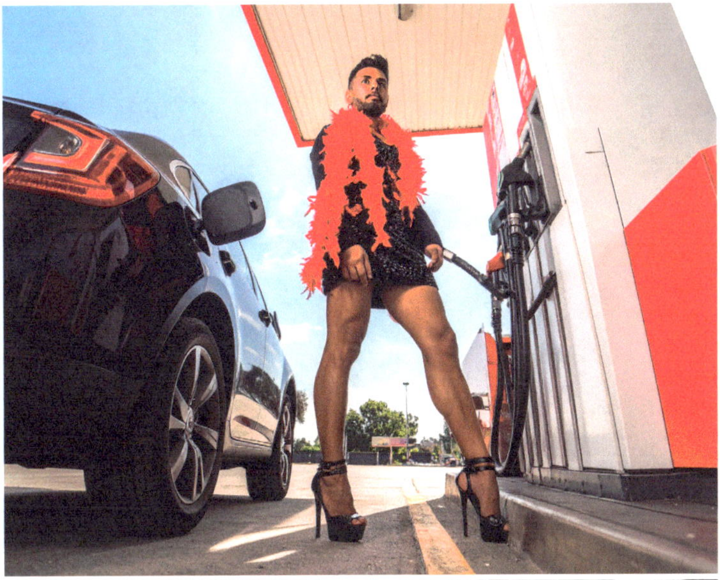

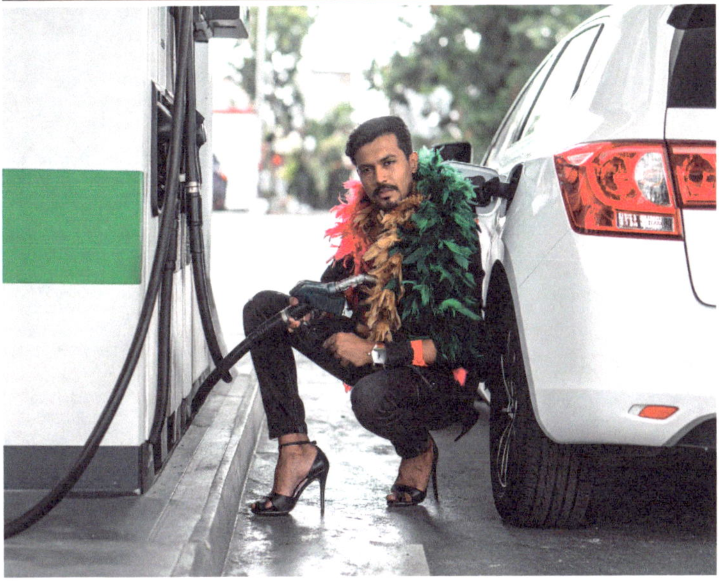

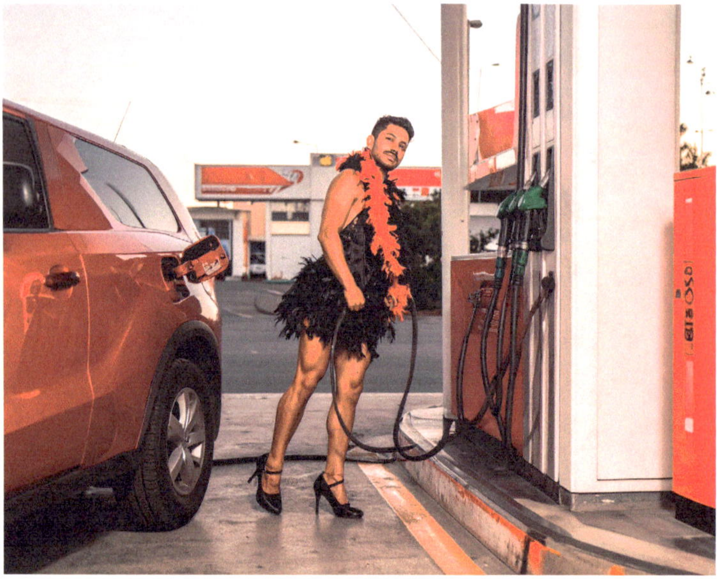

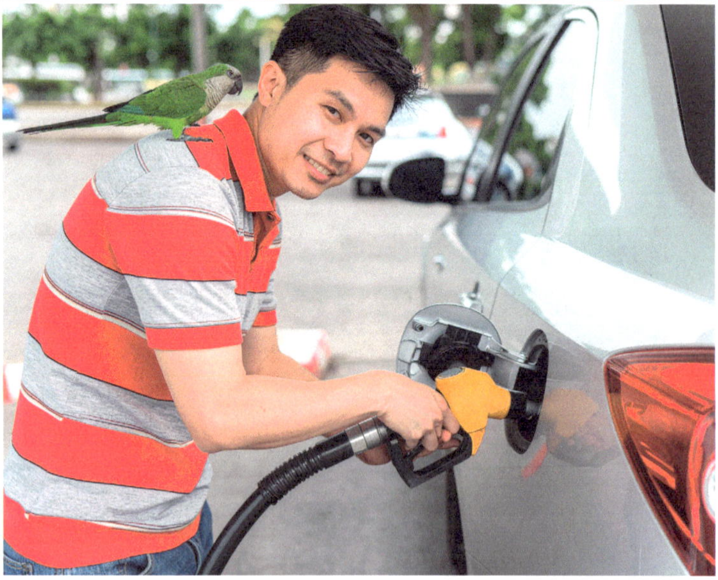

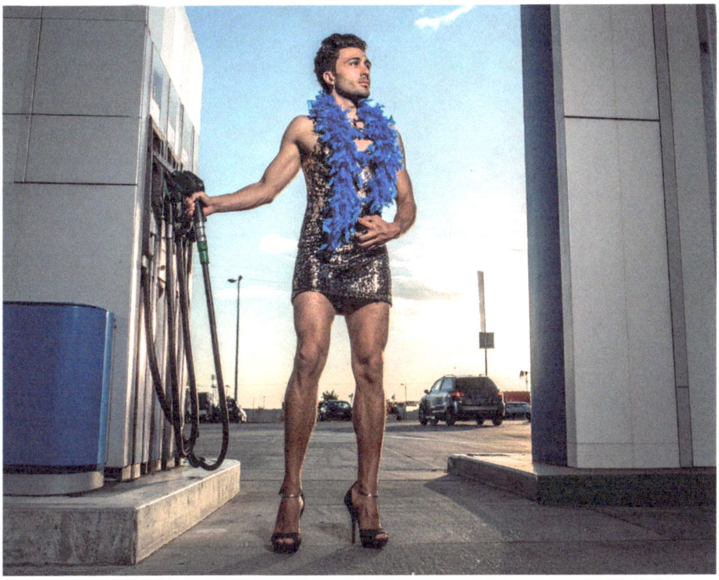

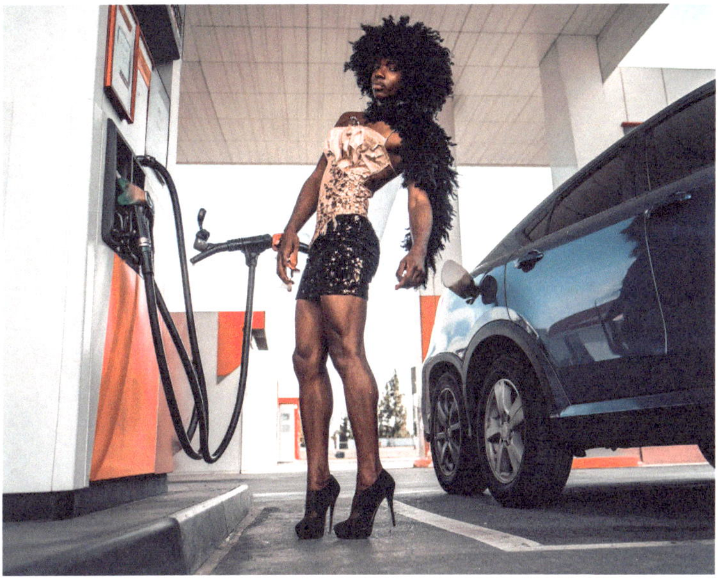

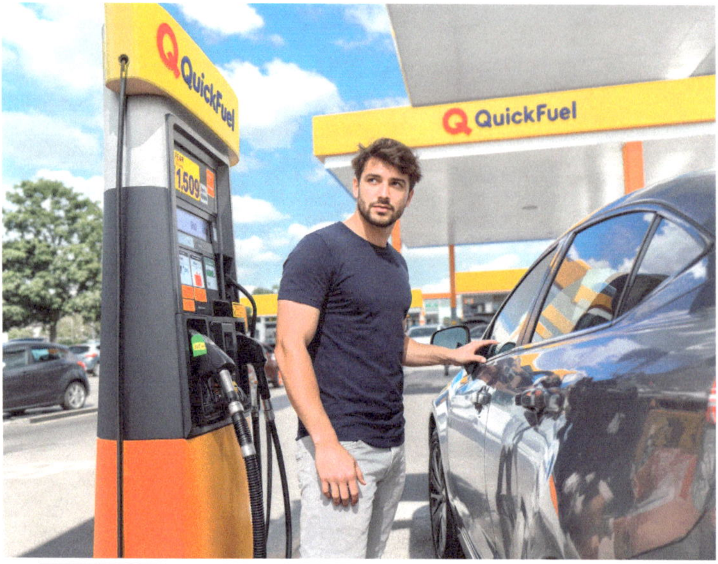
QuickFuel
QuickFuel

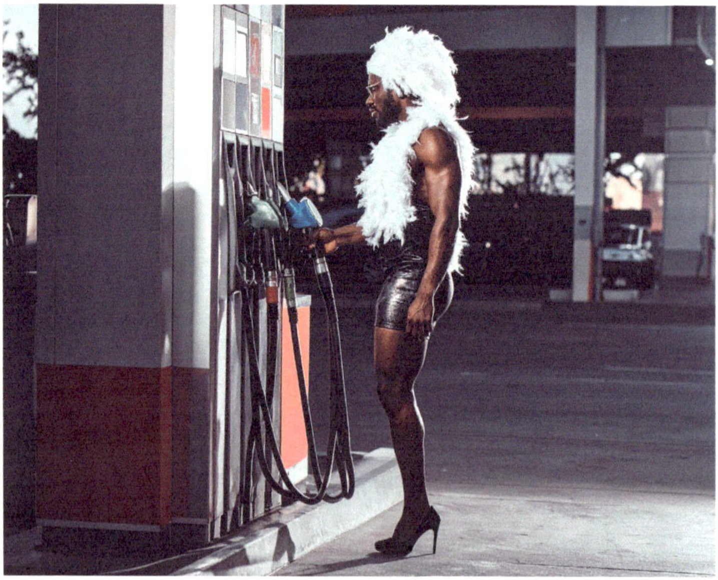

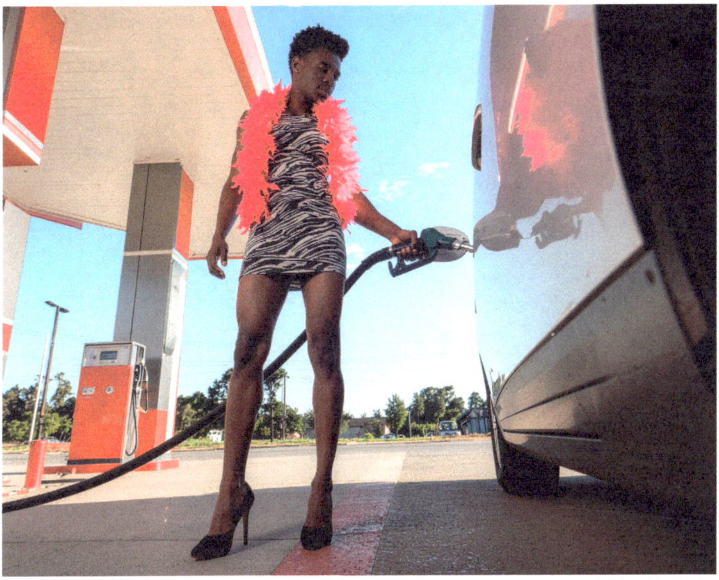

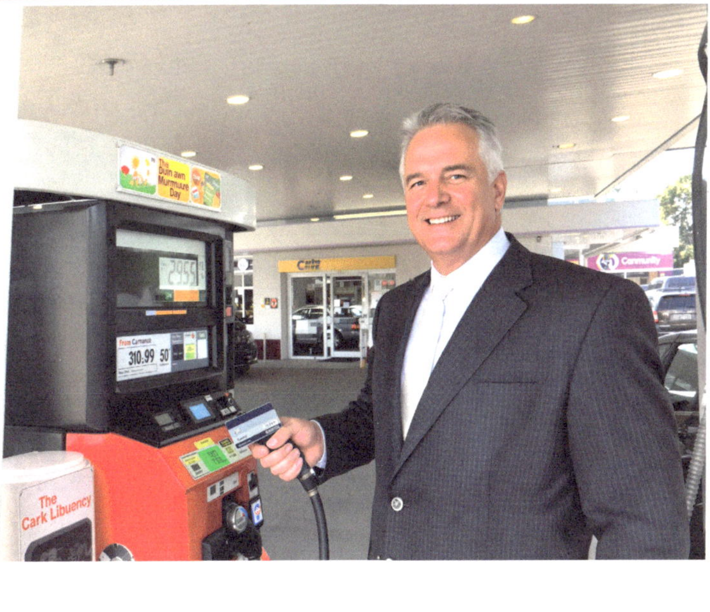

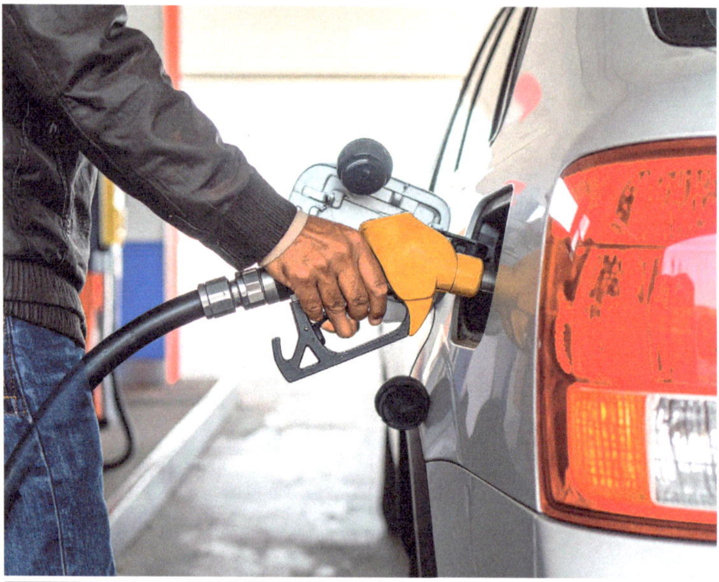

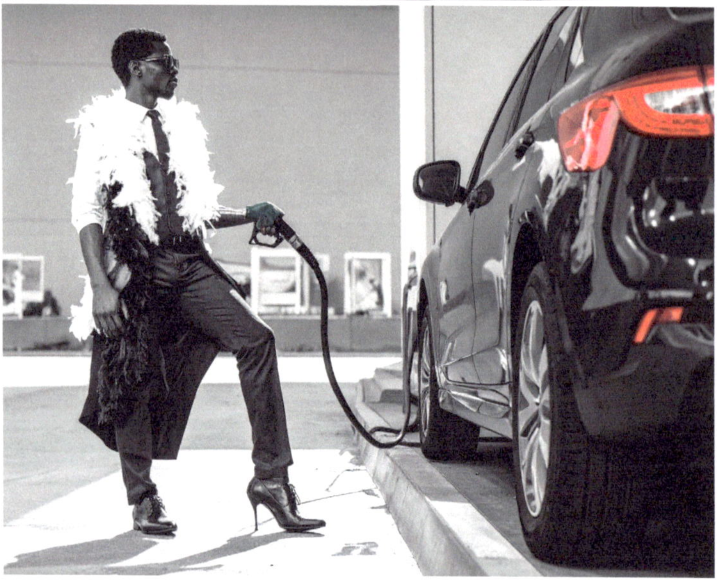

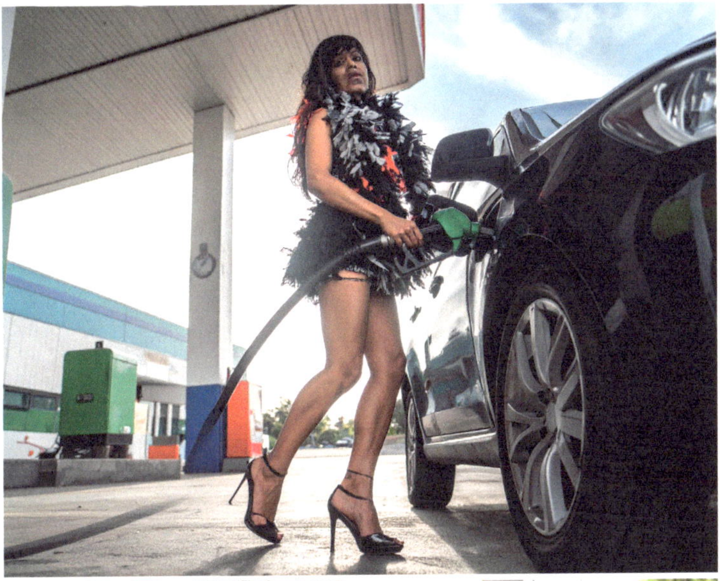

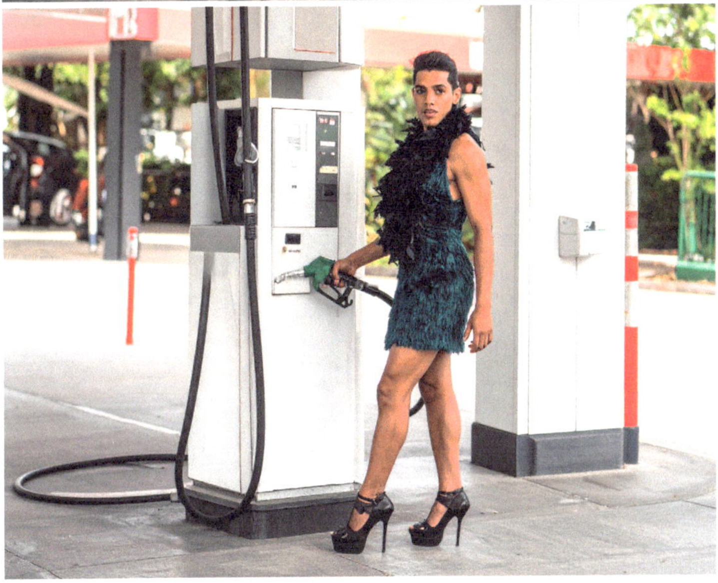

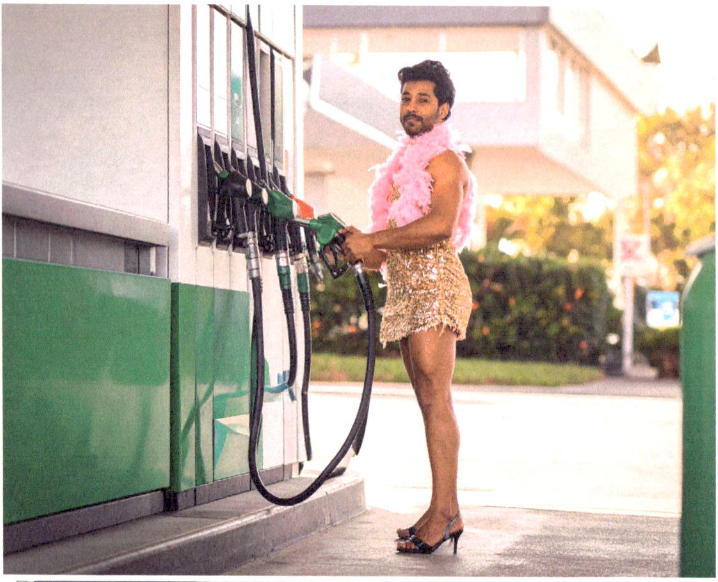

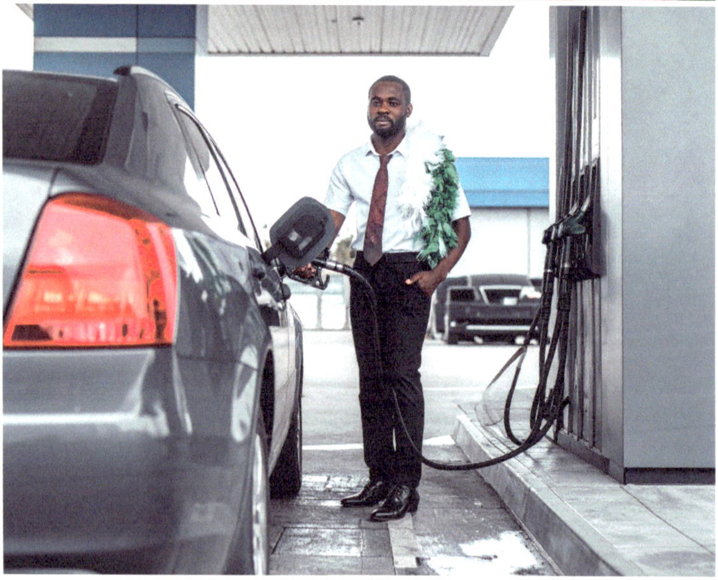

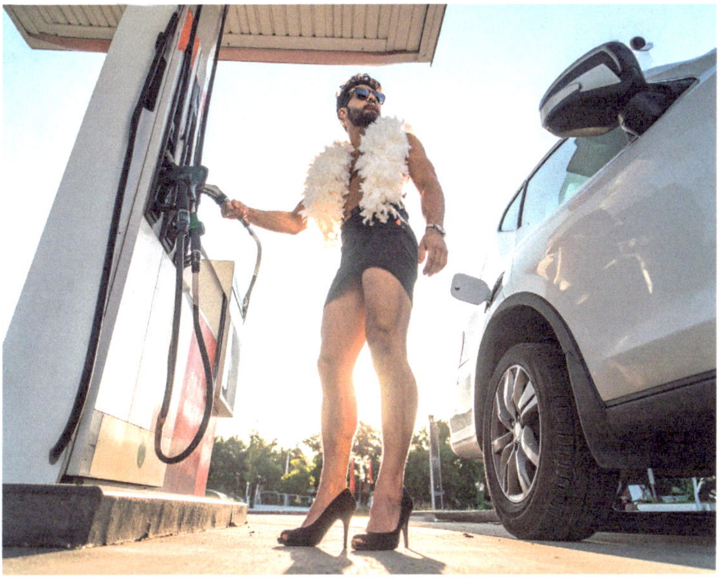

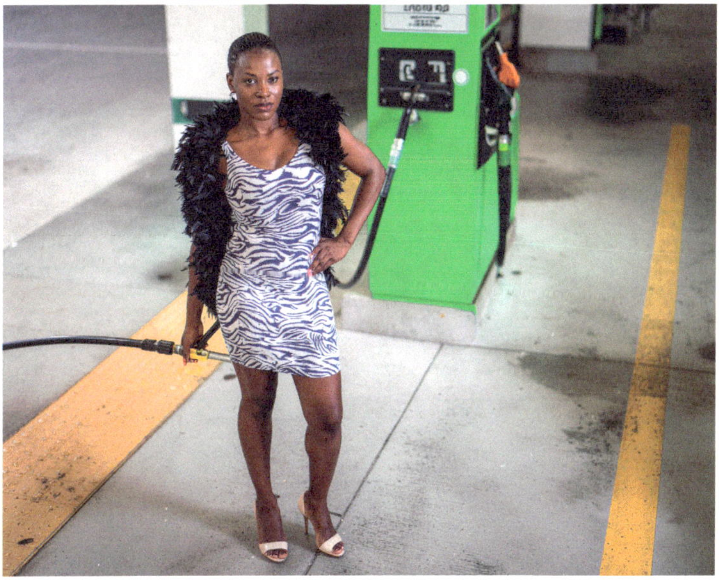

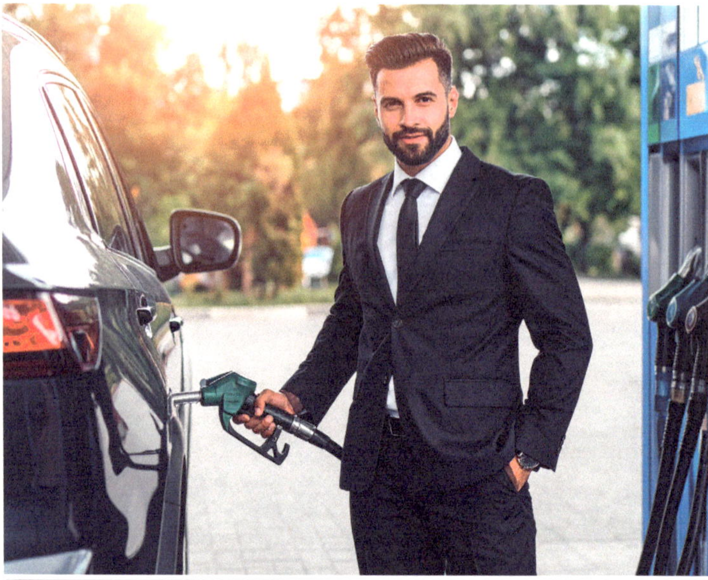

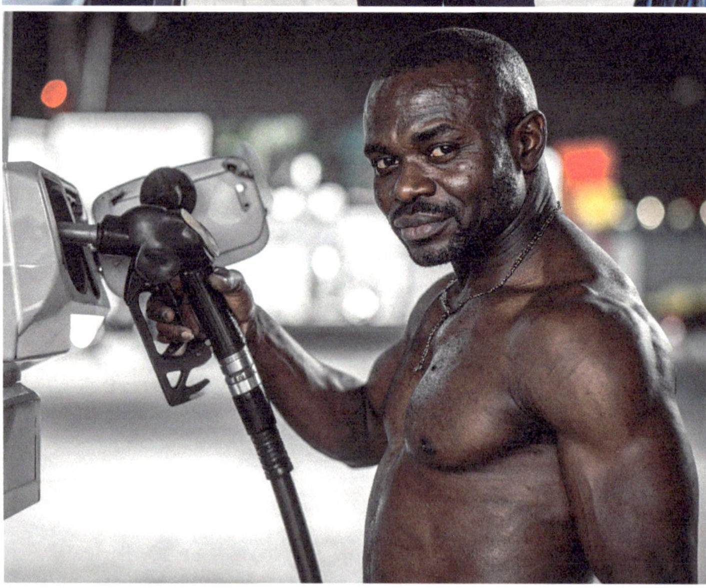

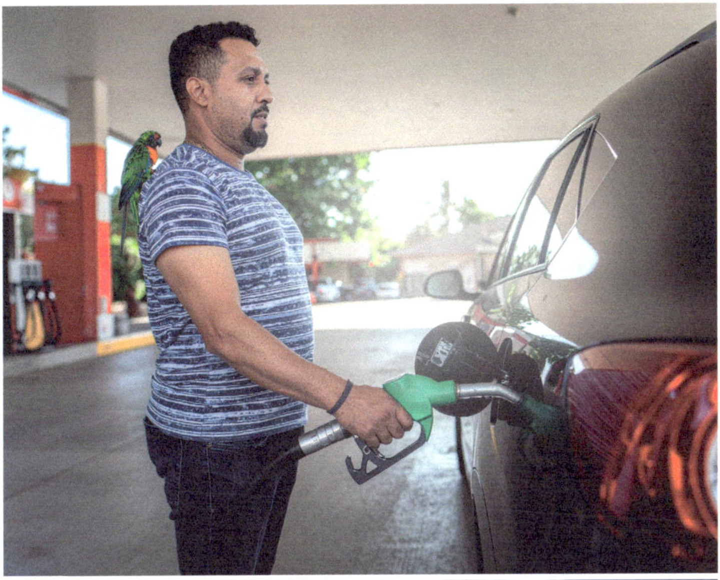

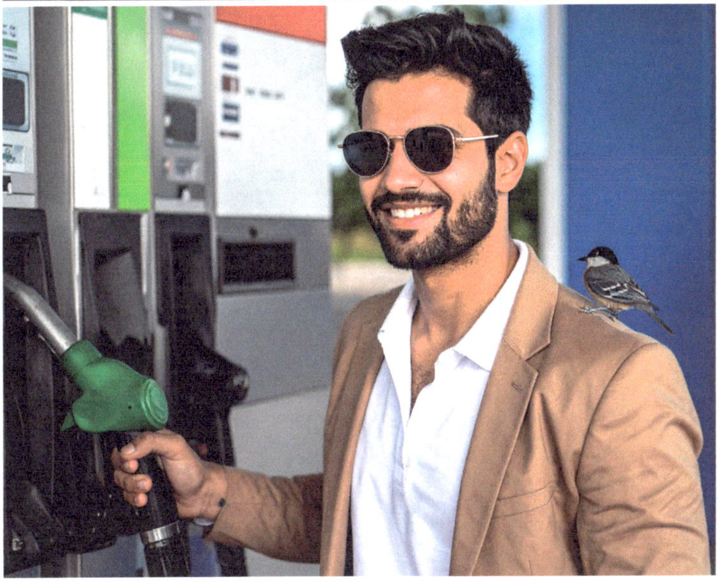

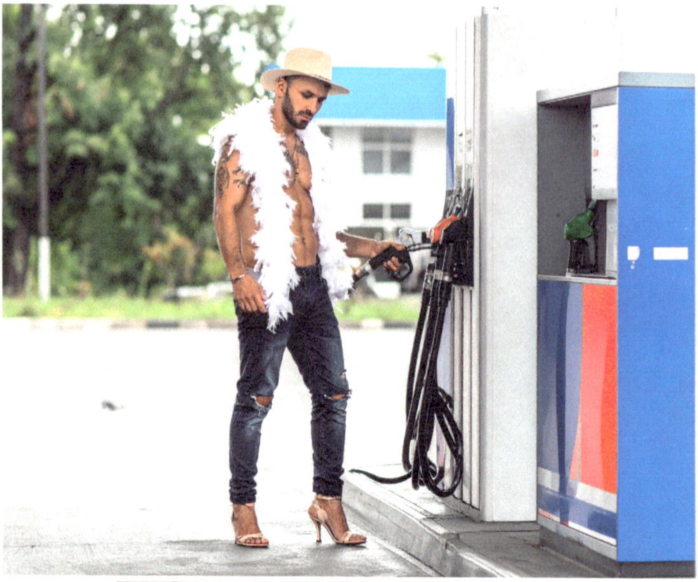

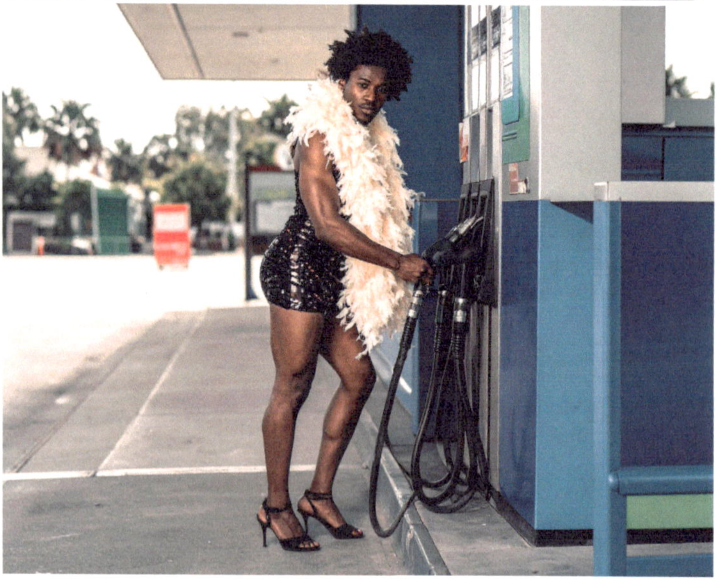

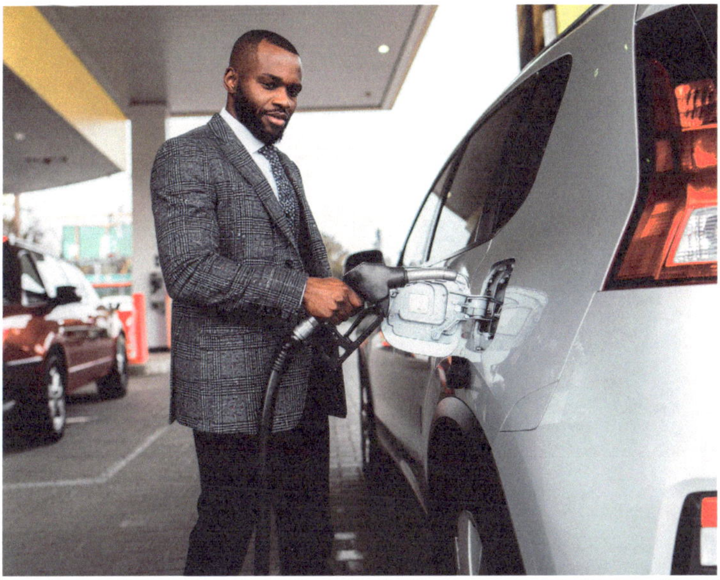

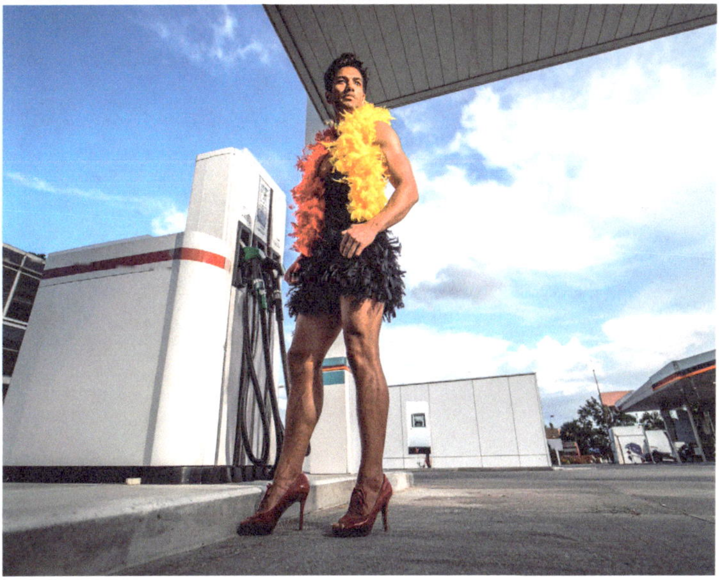

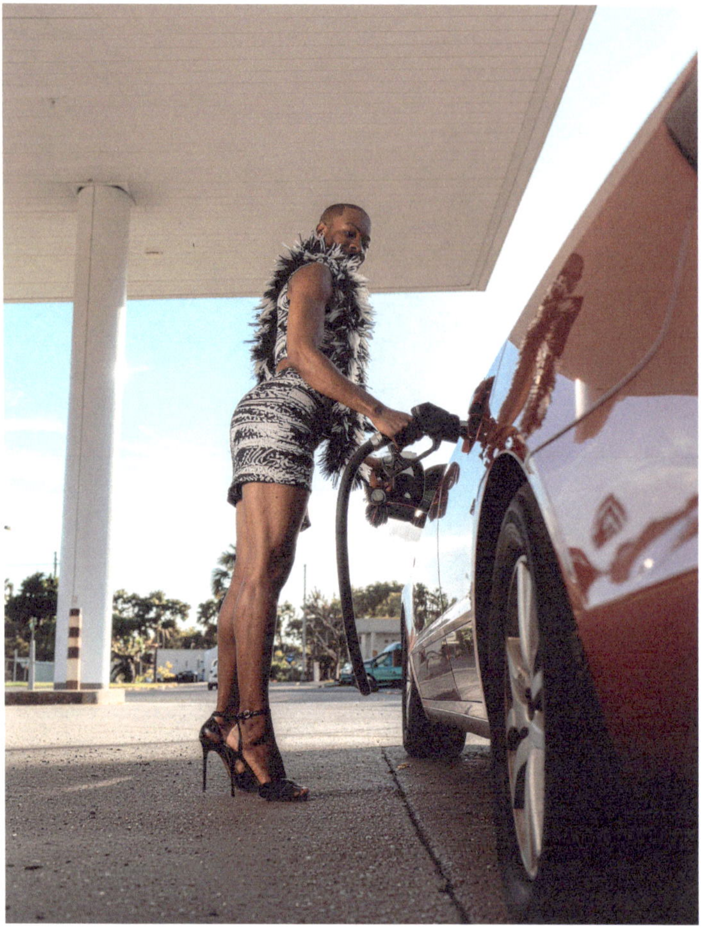

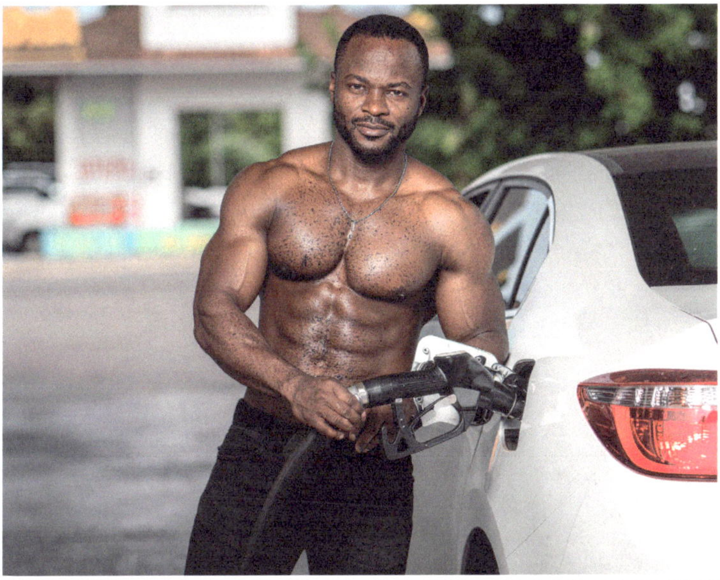

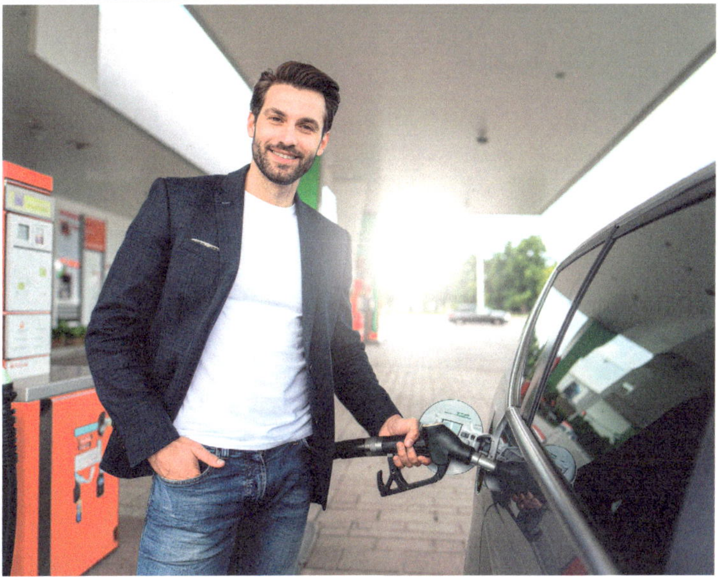

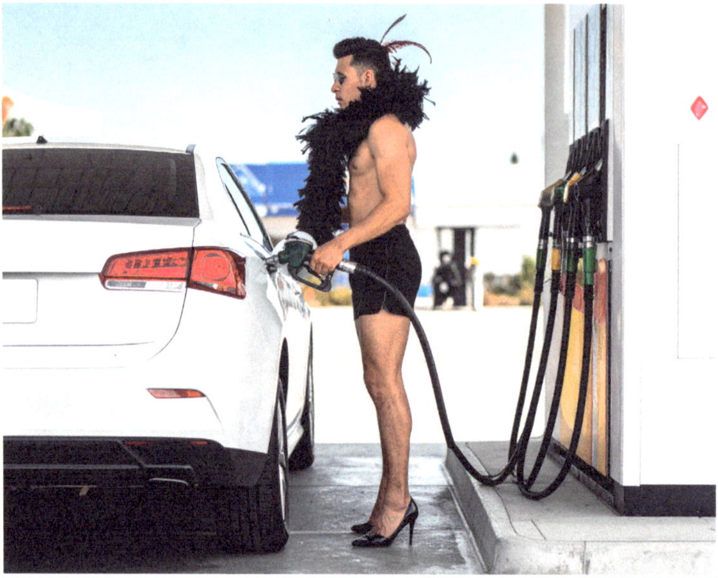

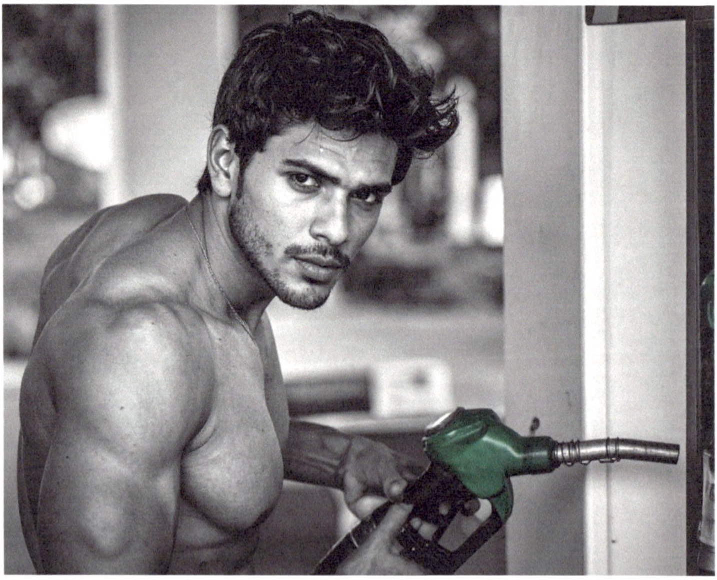

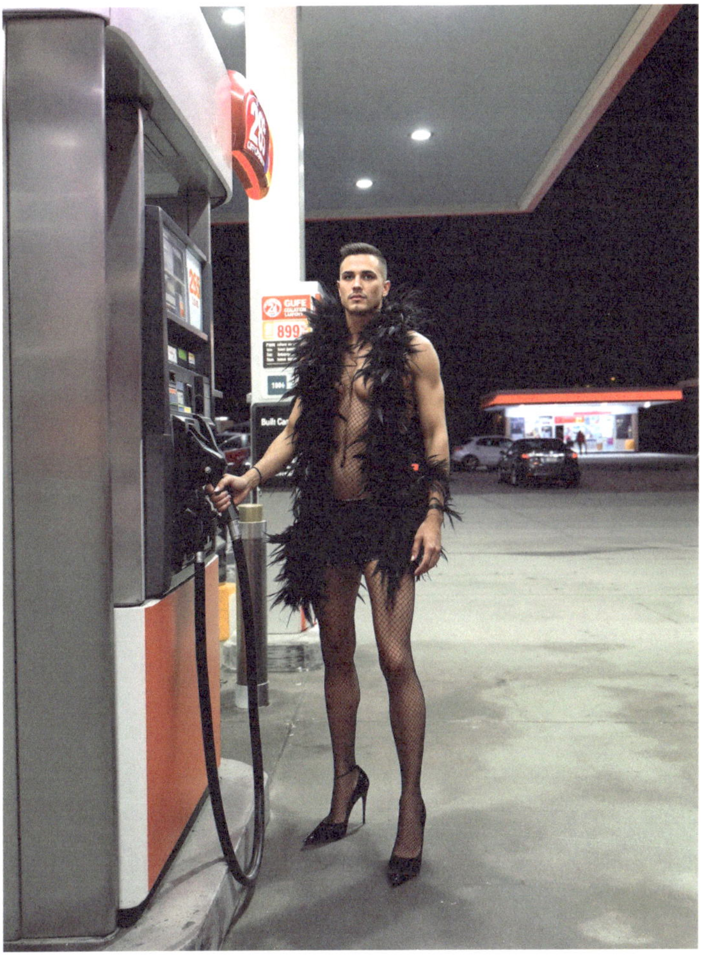

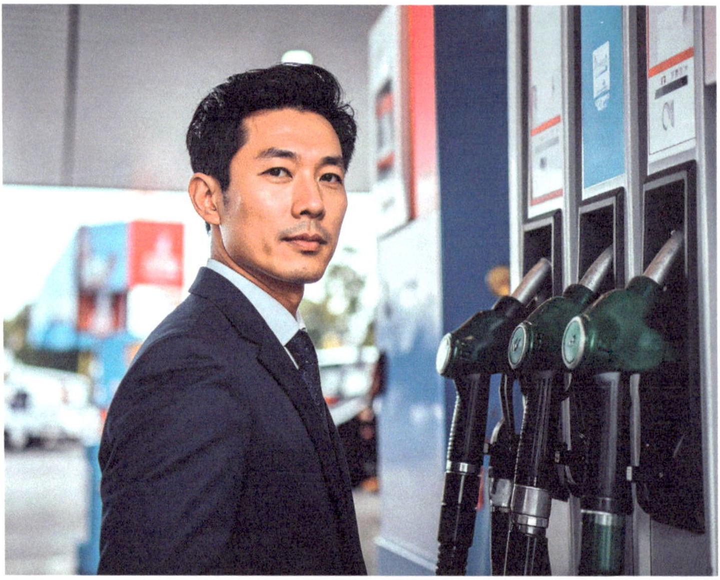

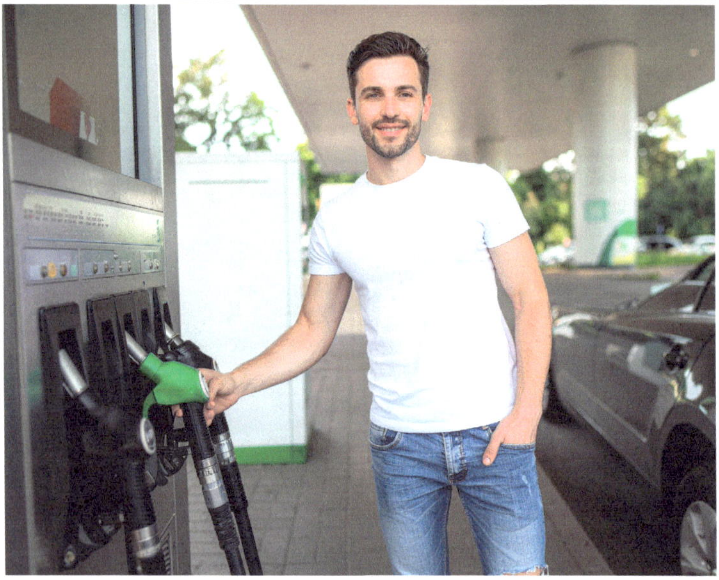

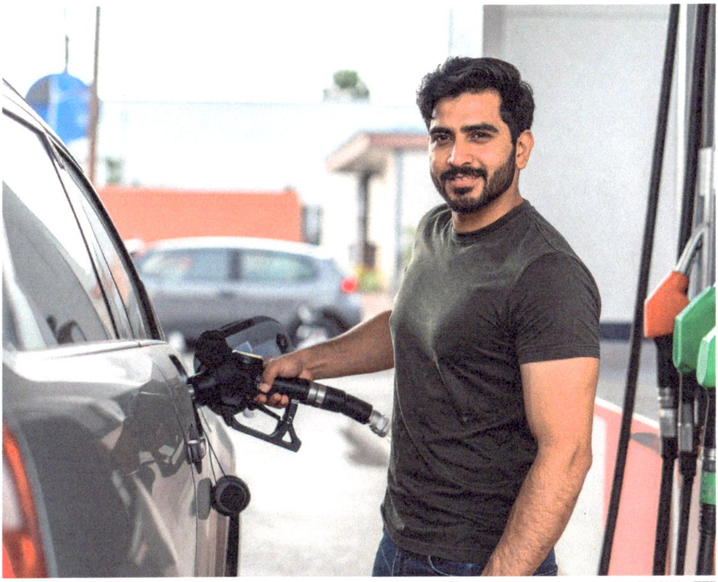

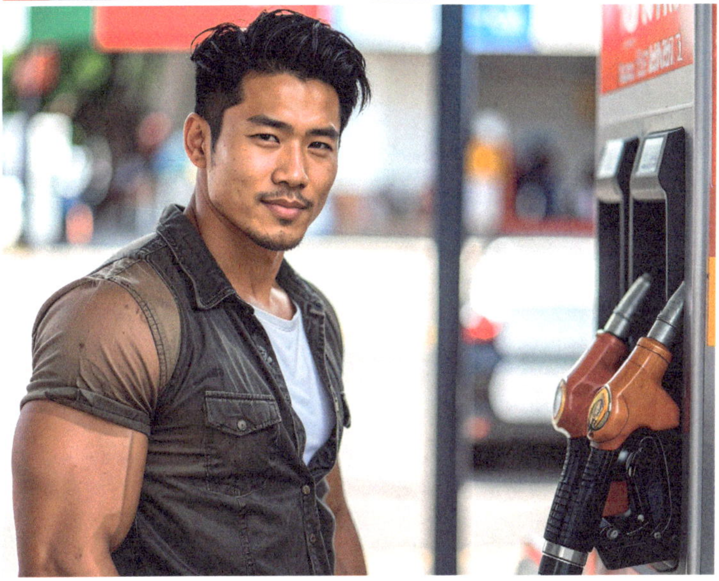

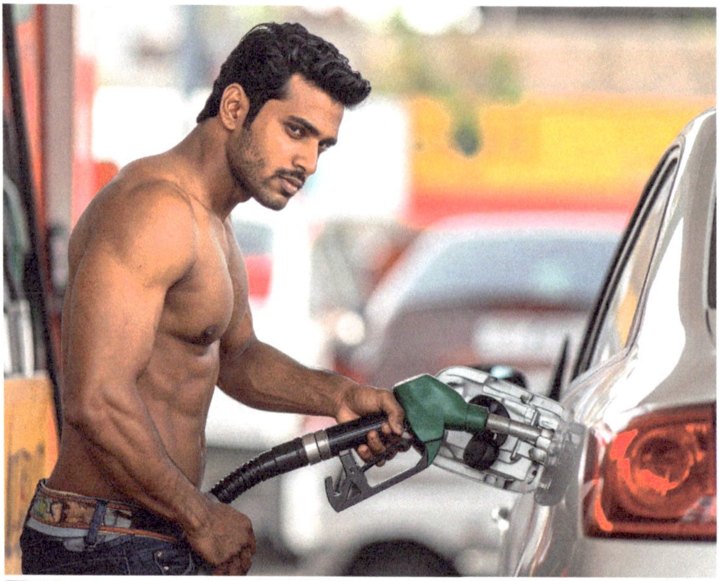

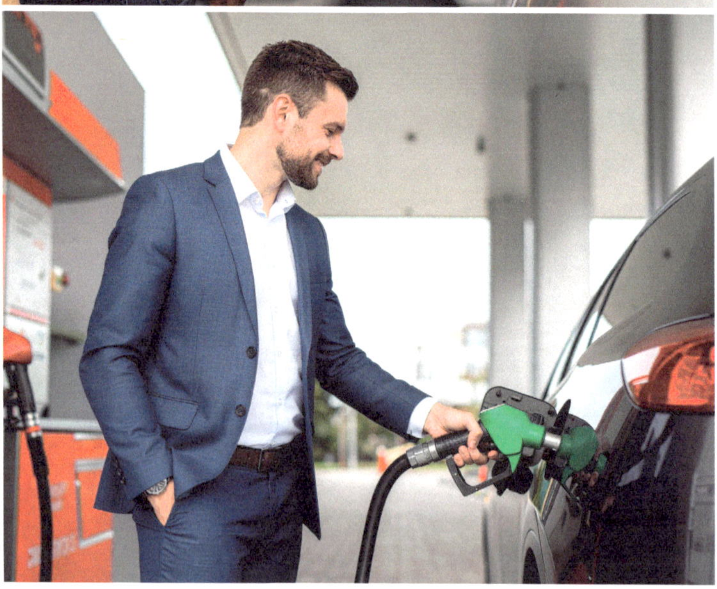

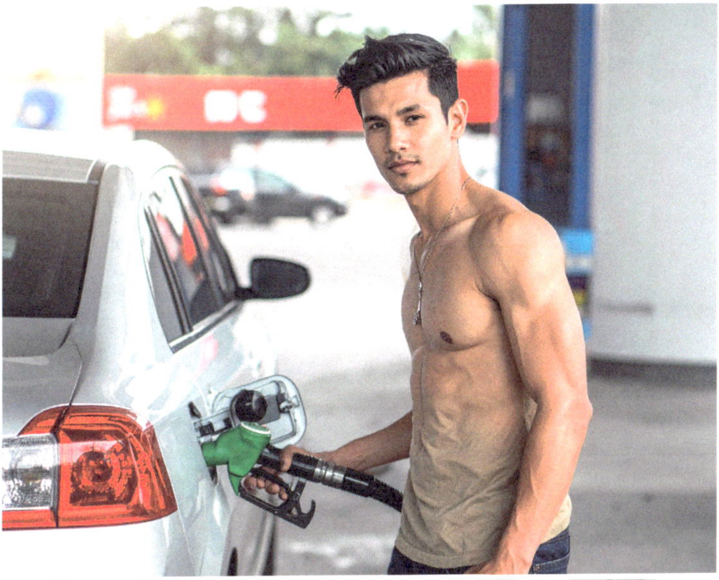

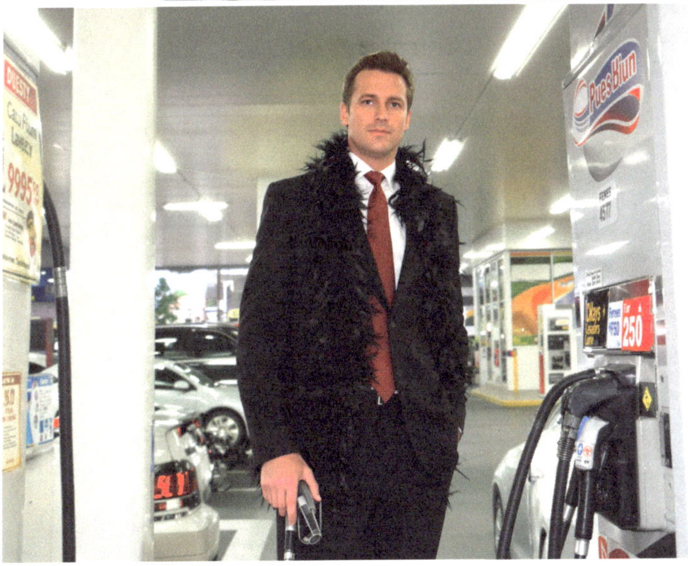

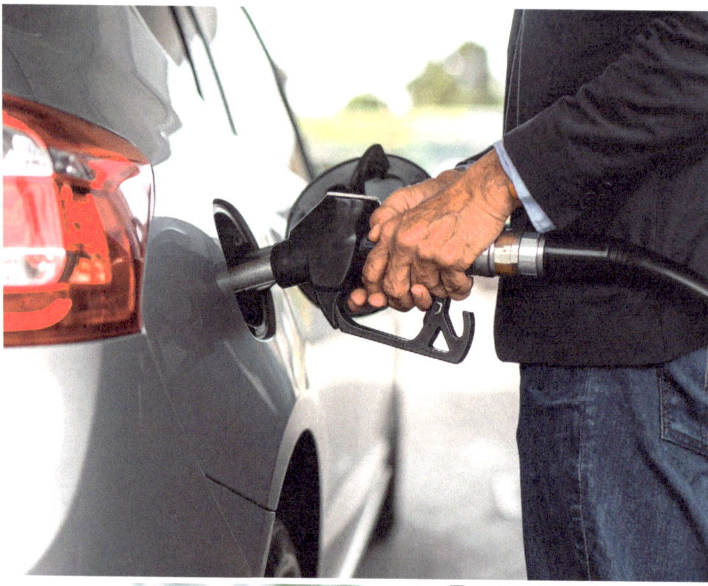

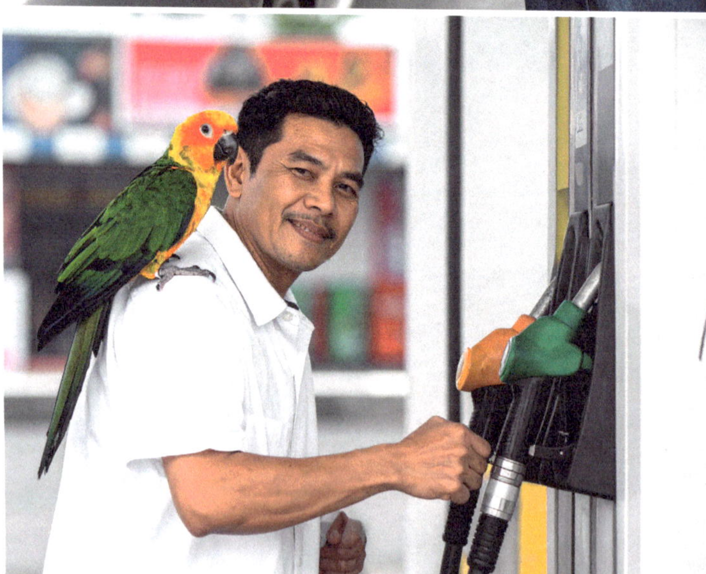

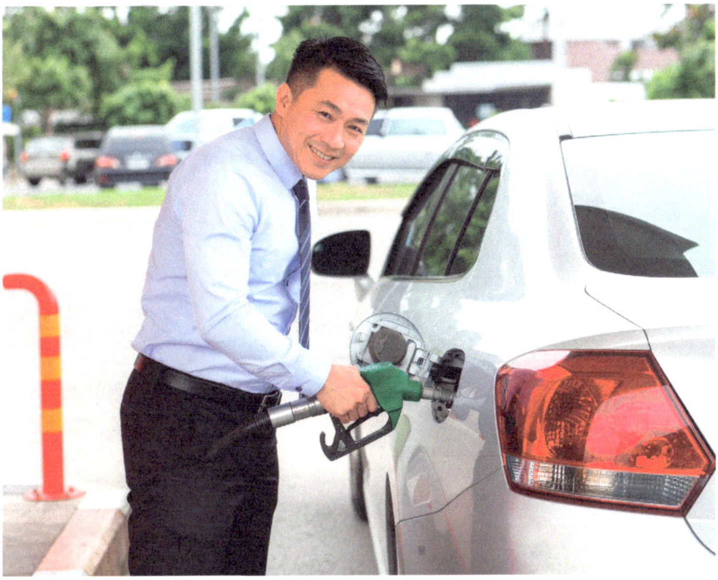

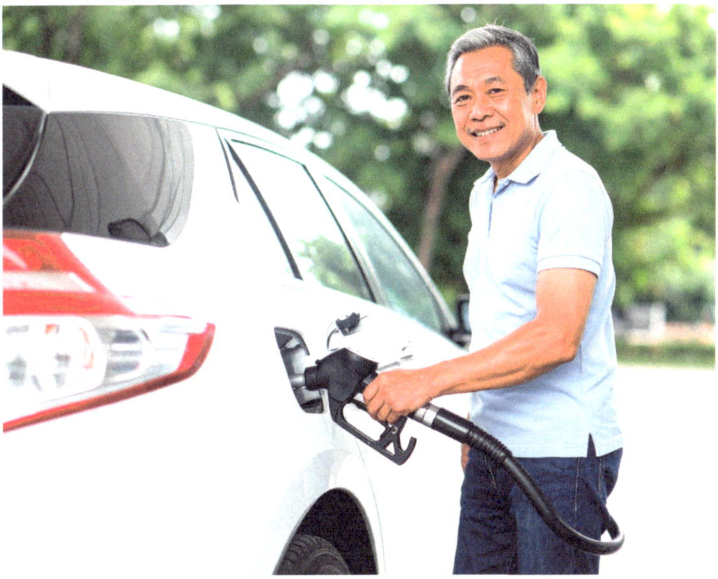

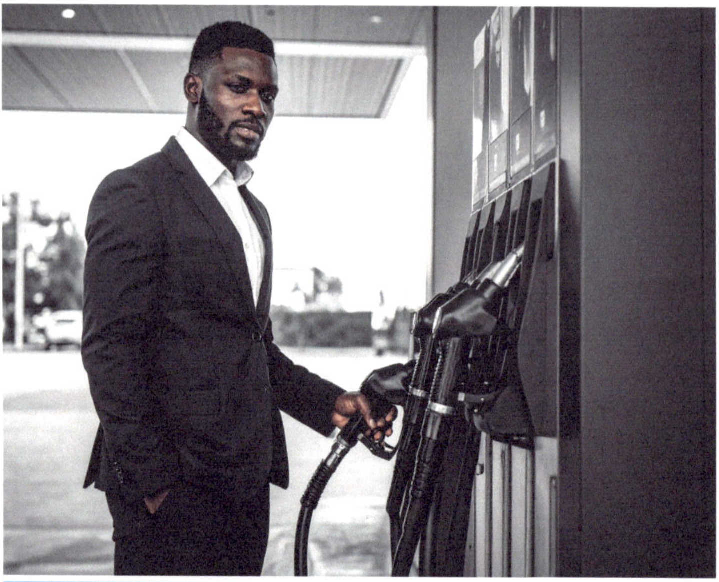

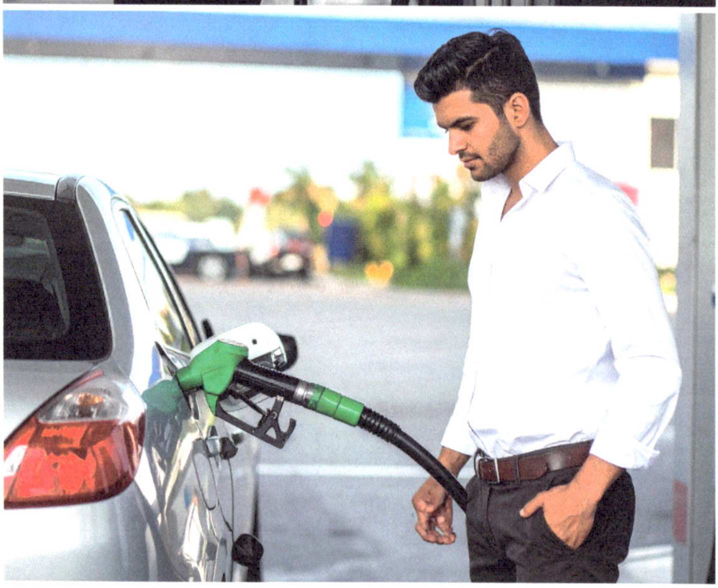

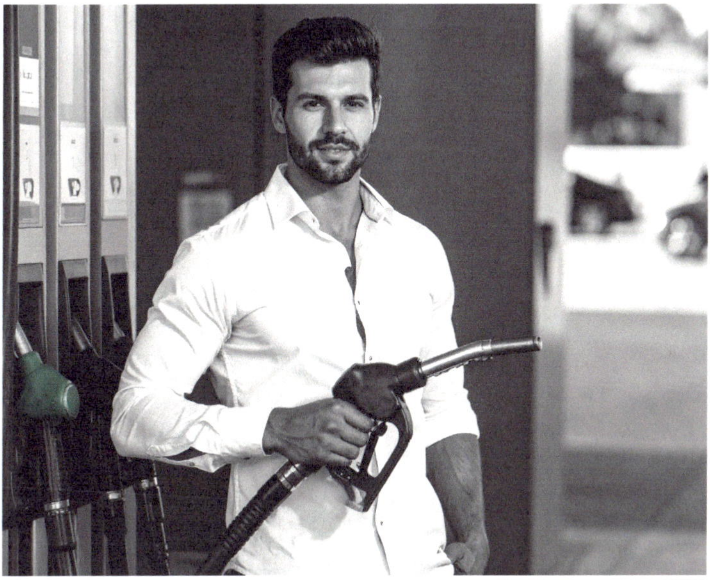

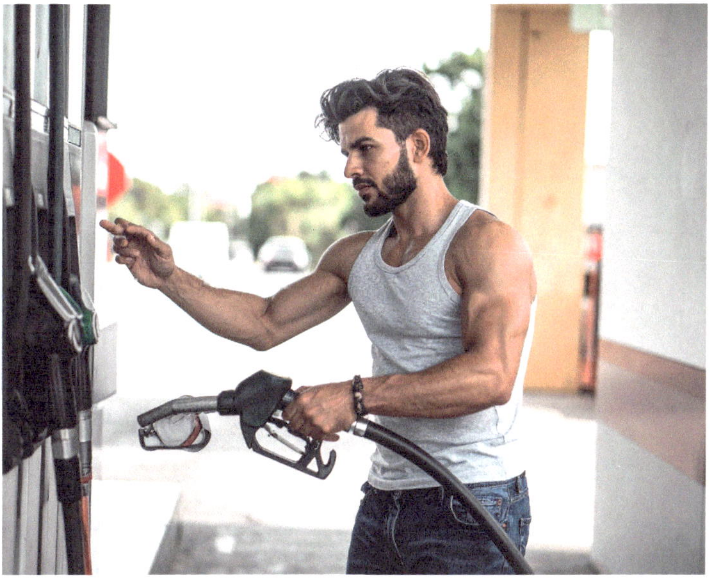

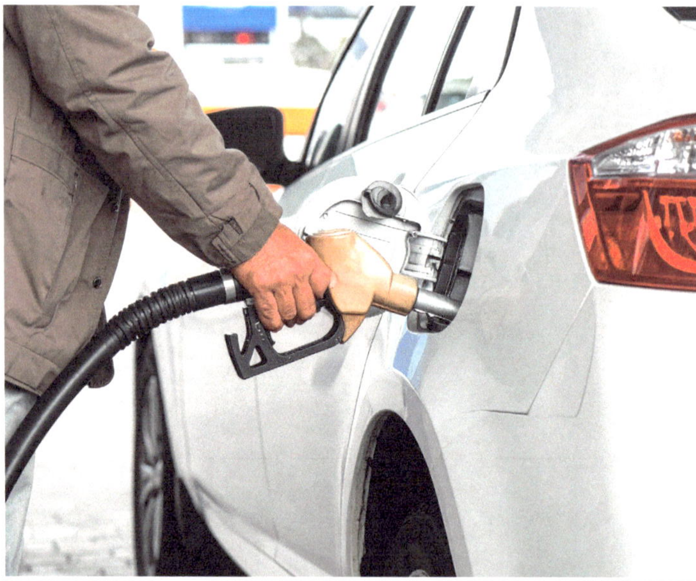

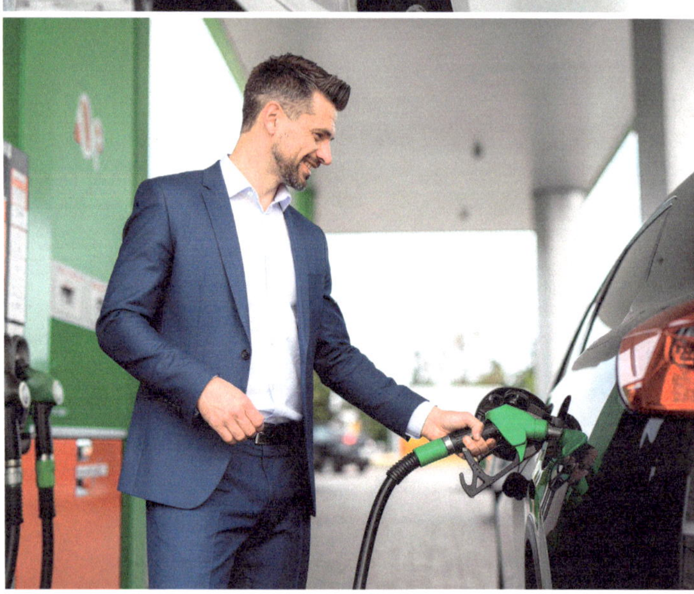

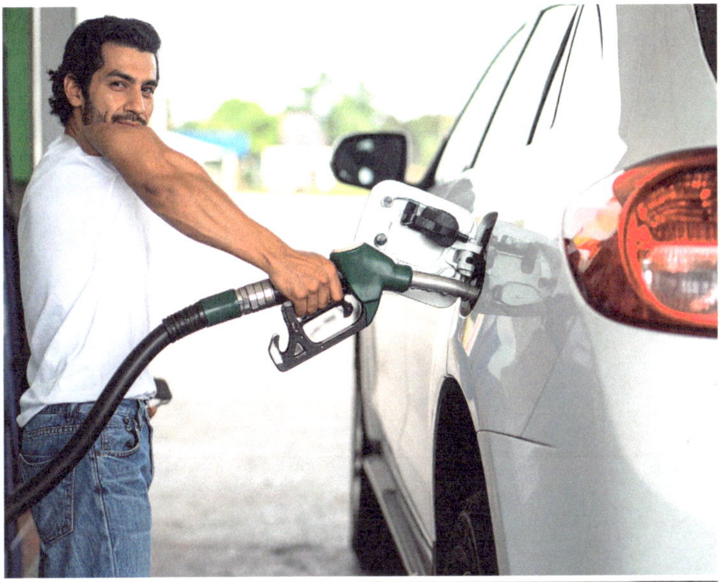

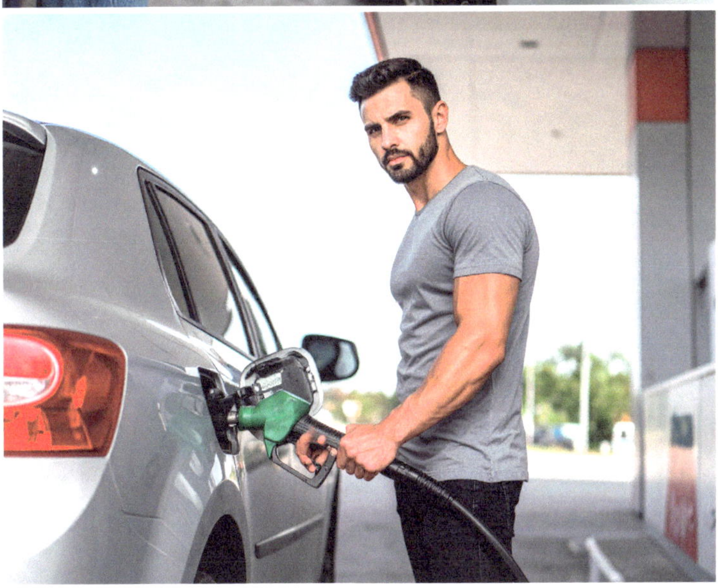

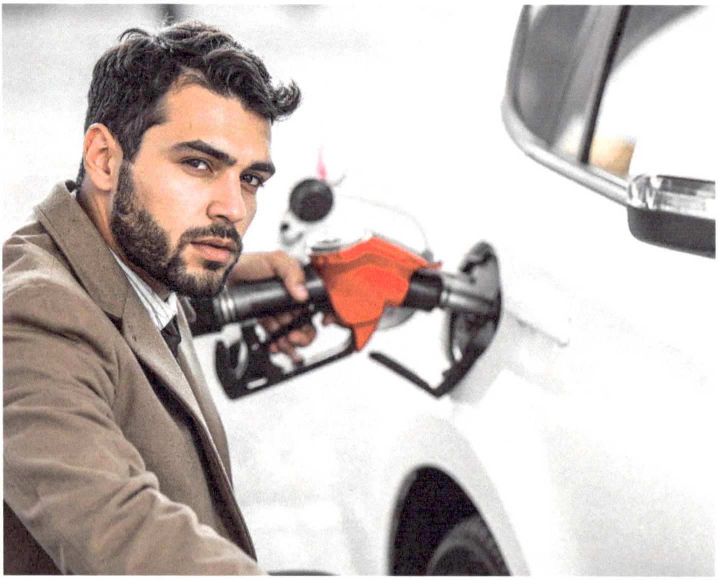

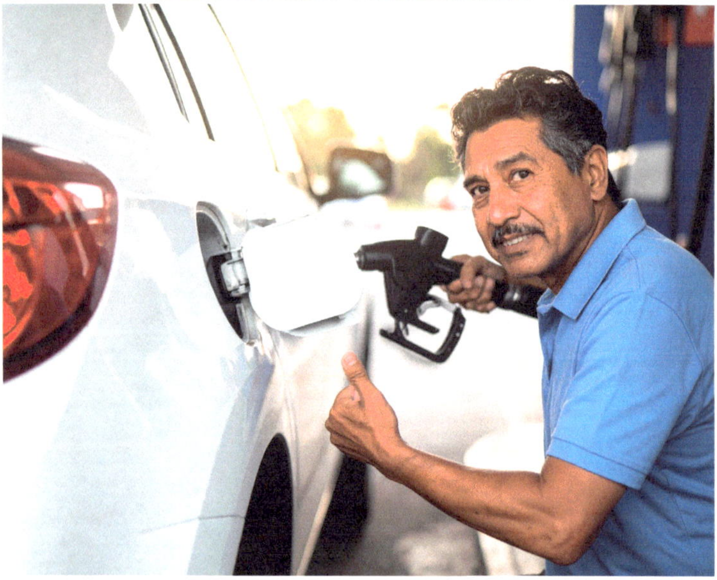

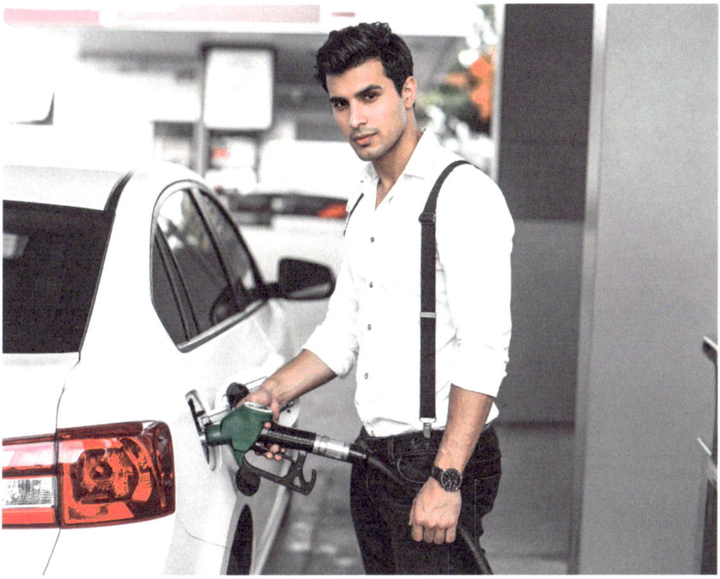

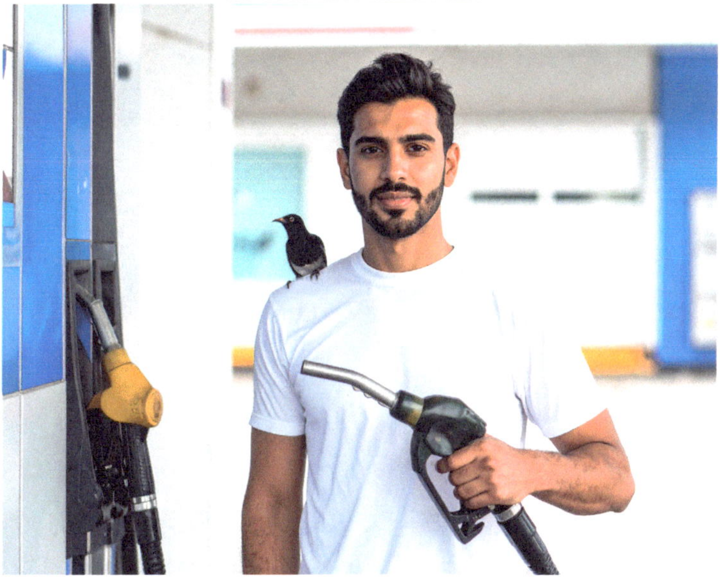

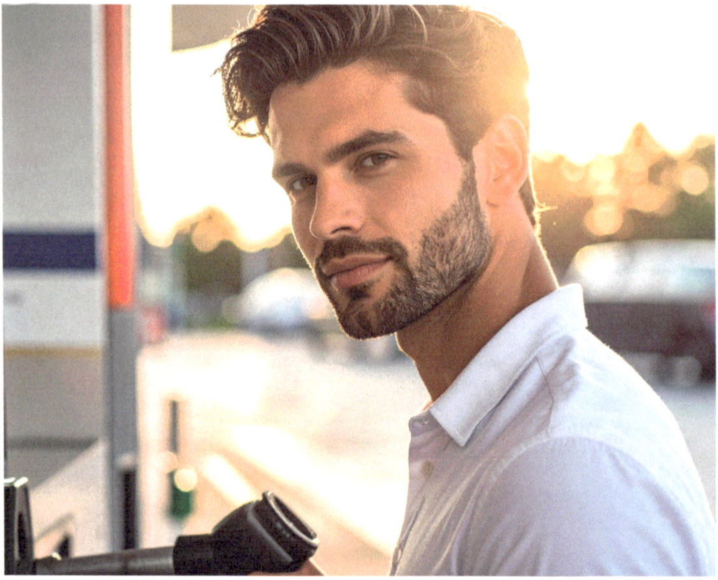

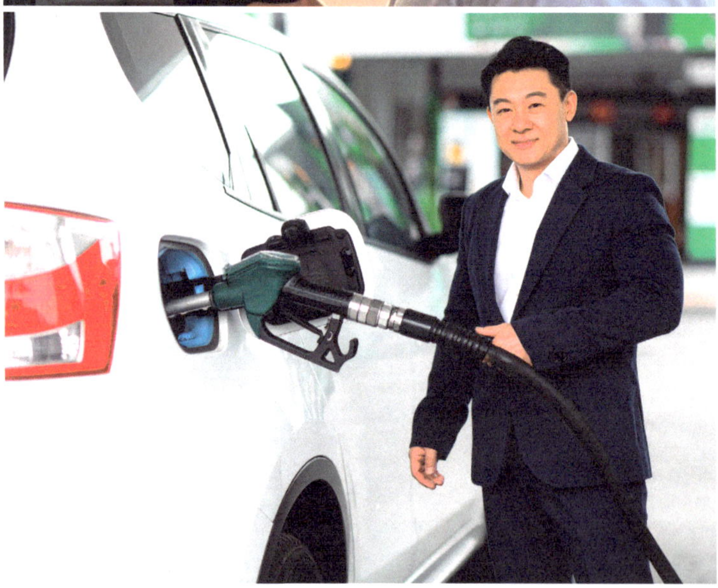

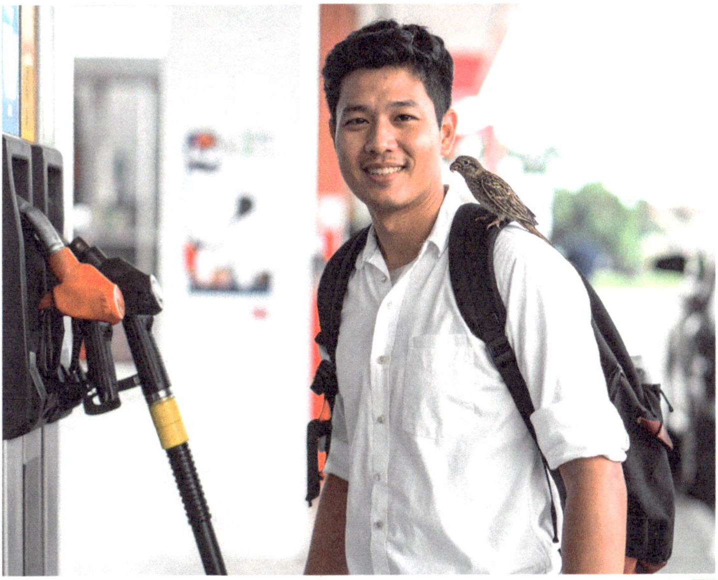

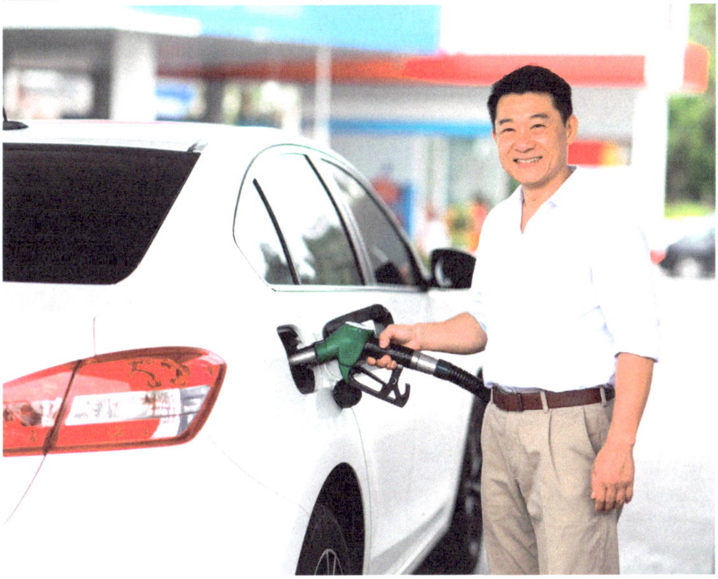

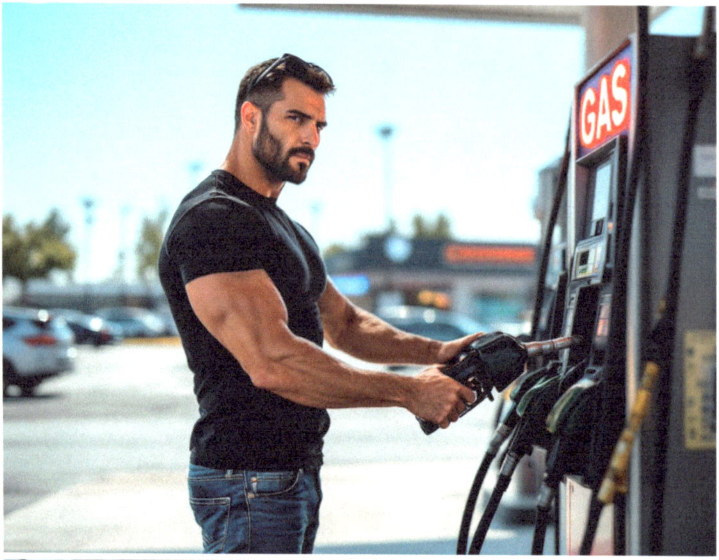
GAS

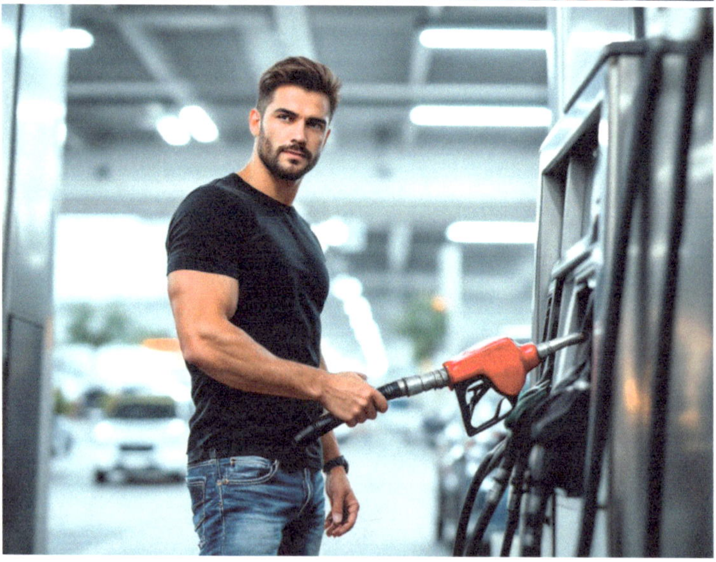

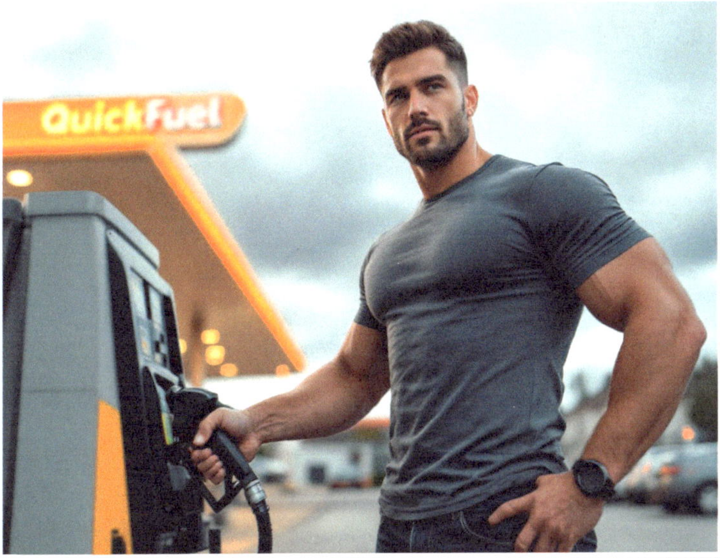
QuickFuel
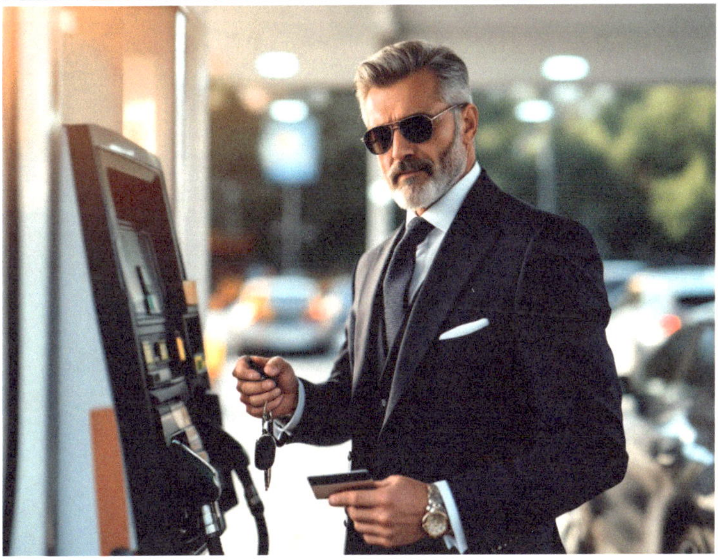

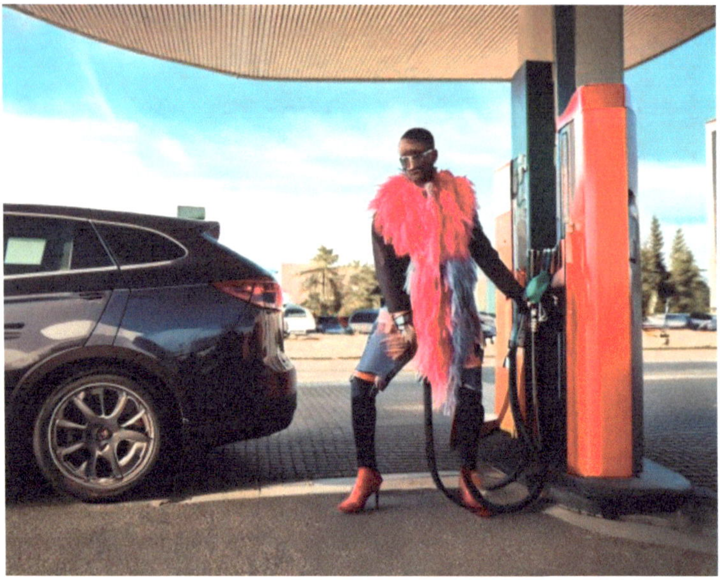

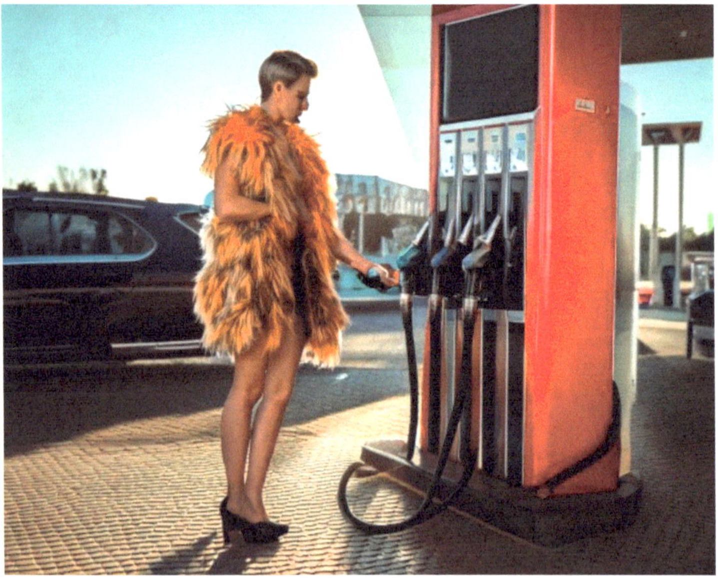

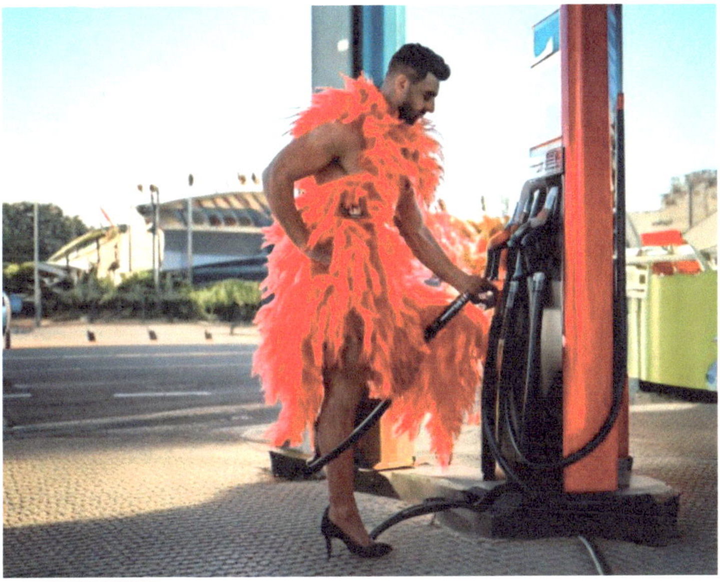

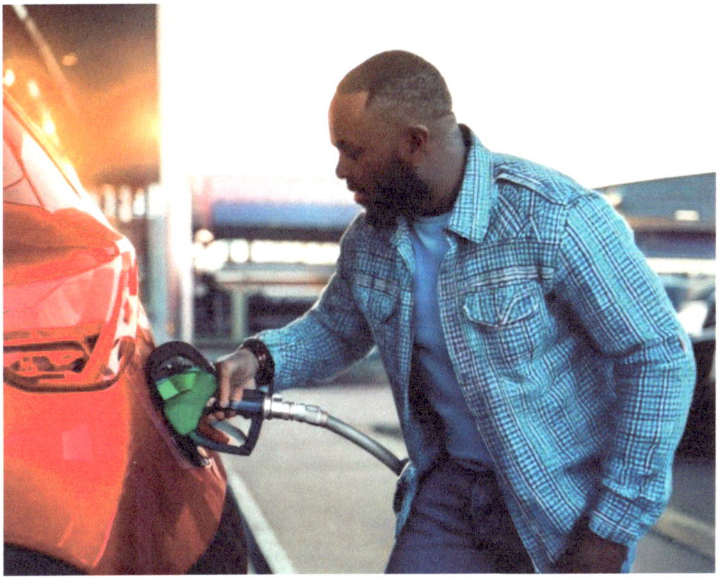

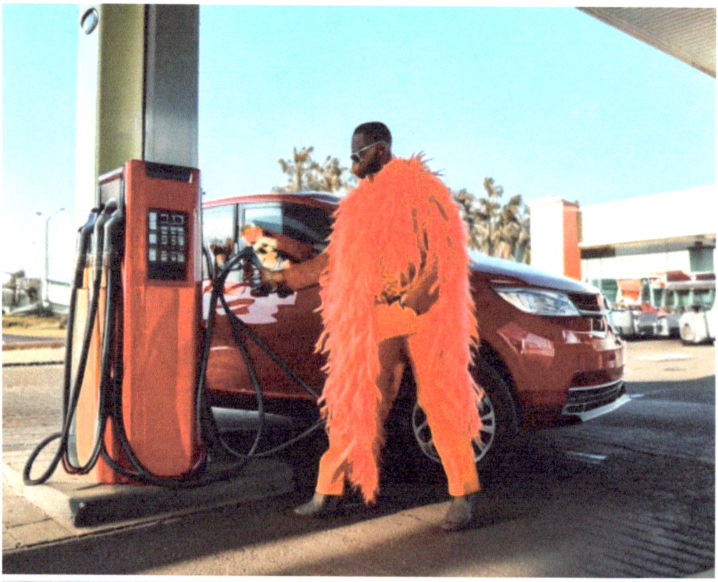

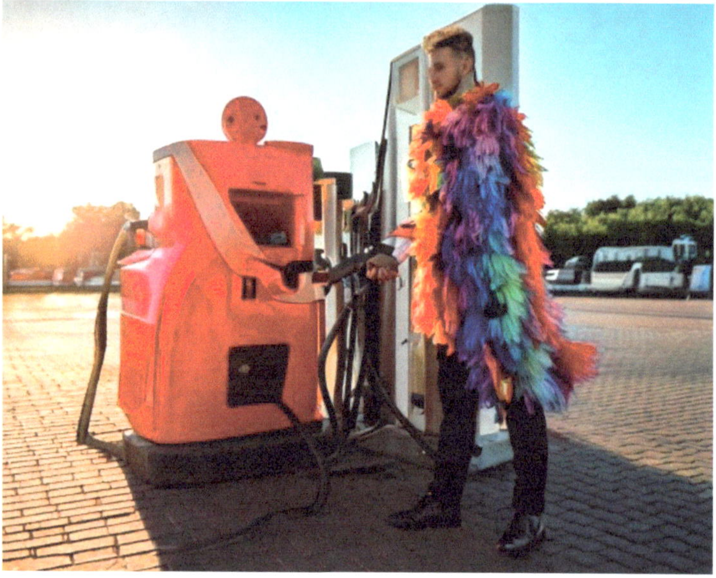

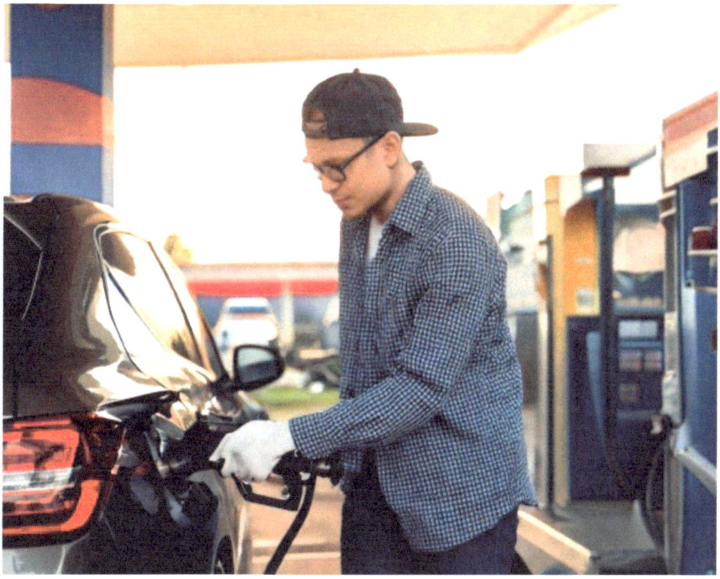

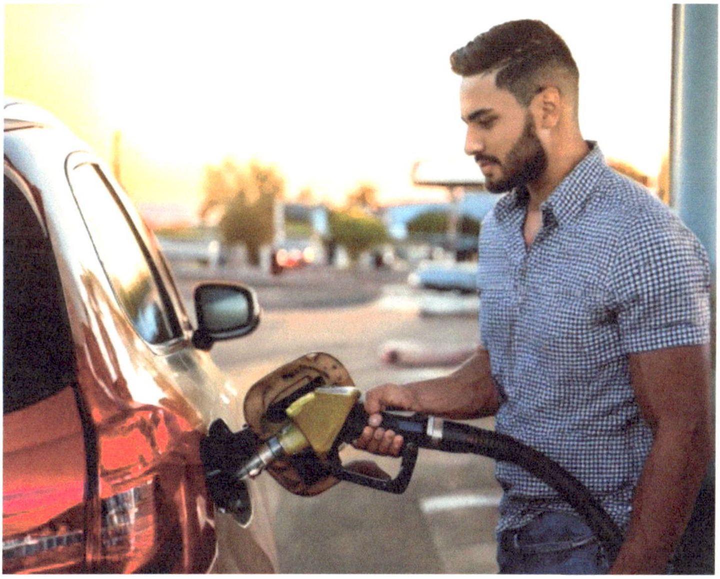

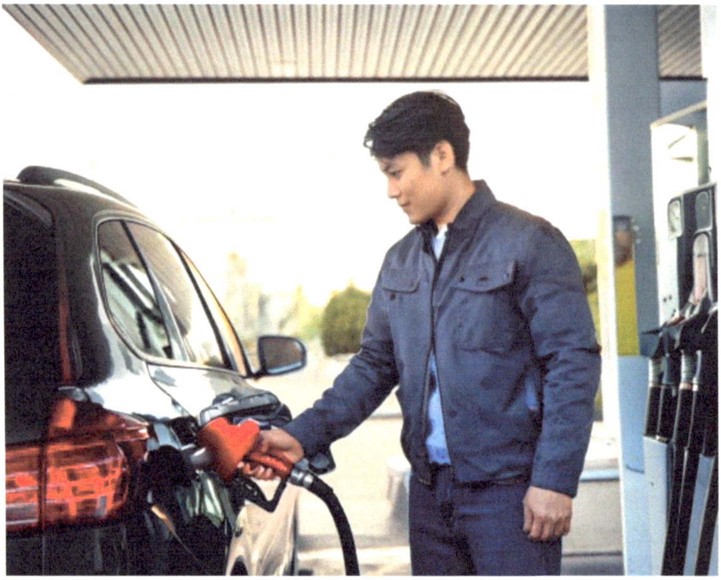

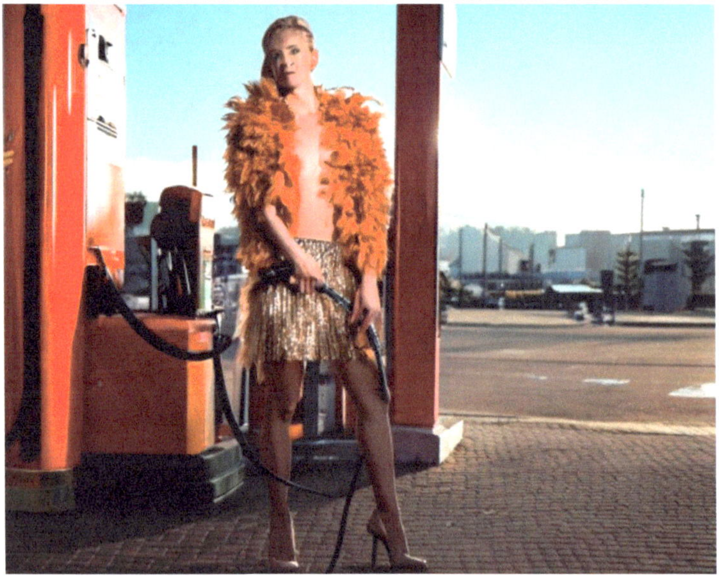

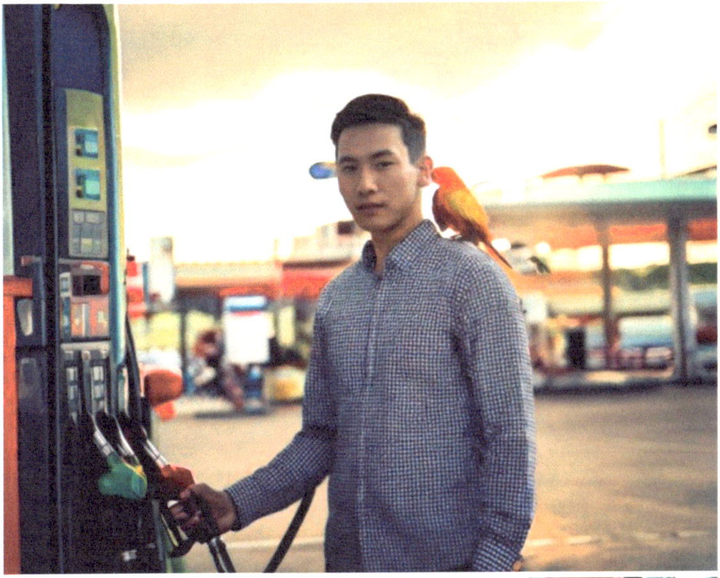

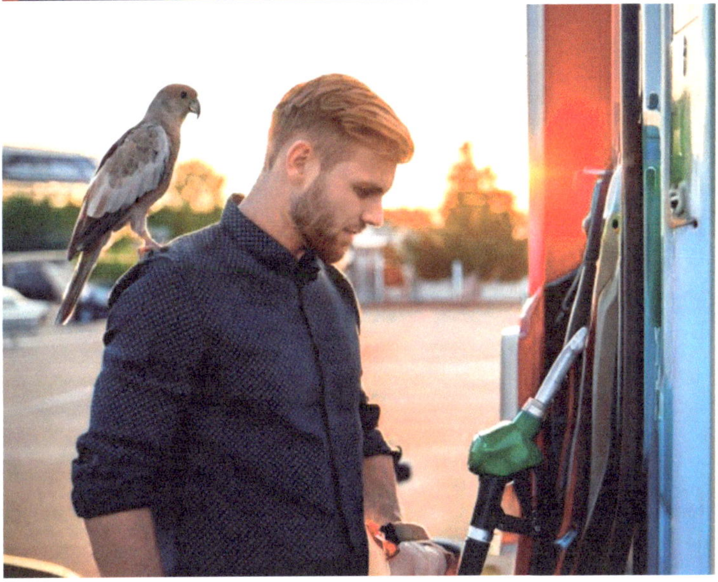

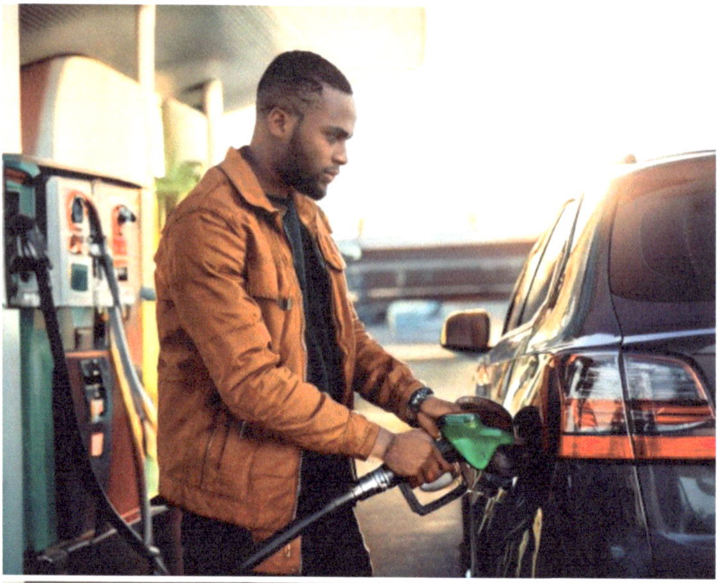

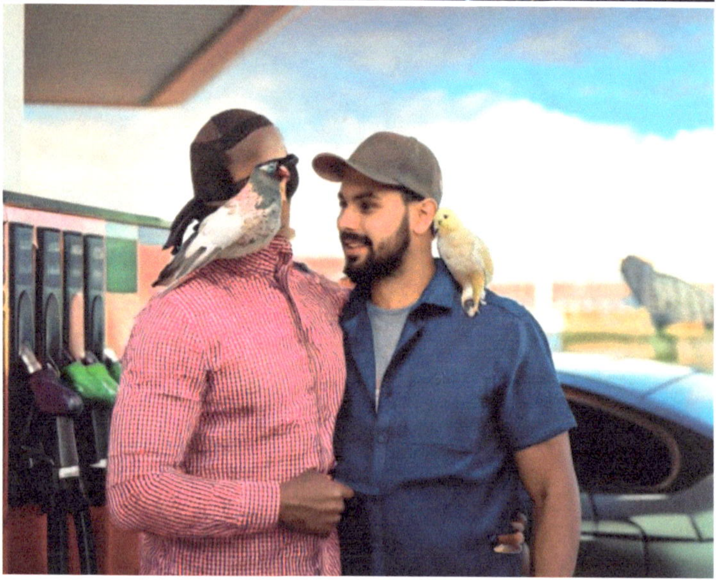

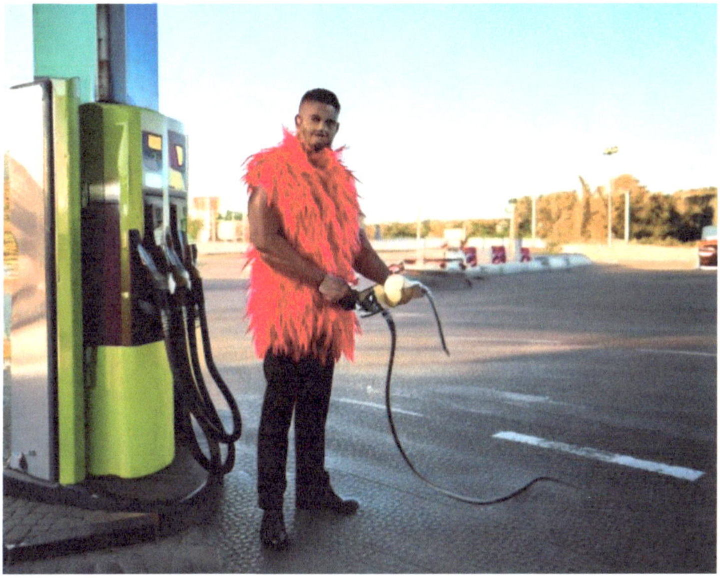

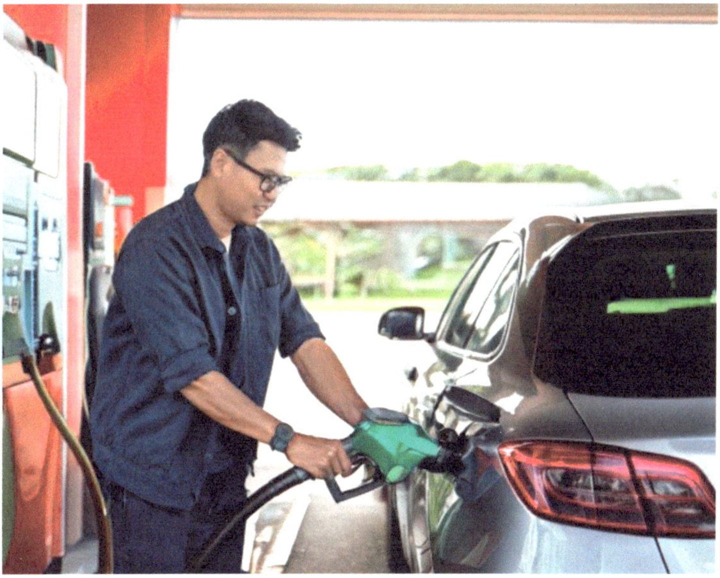

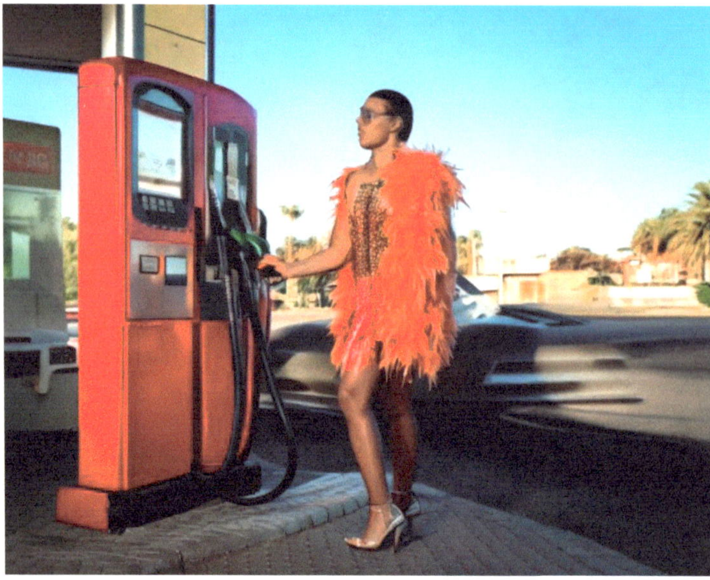

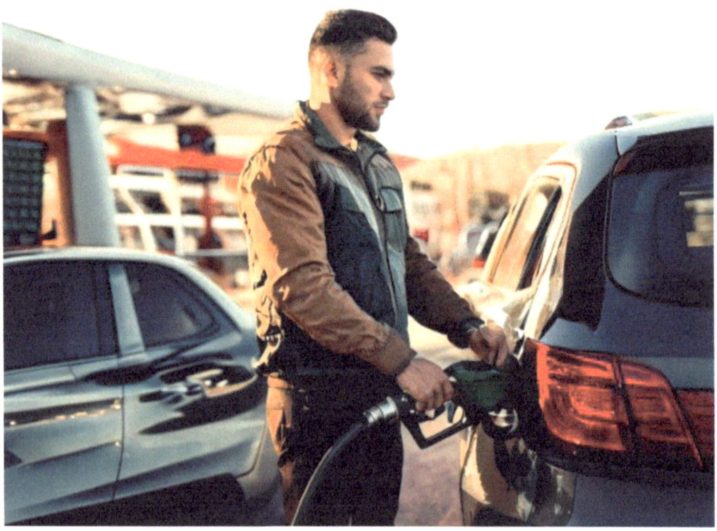

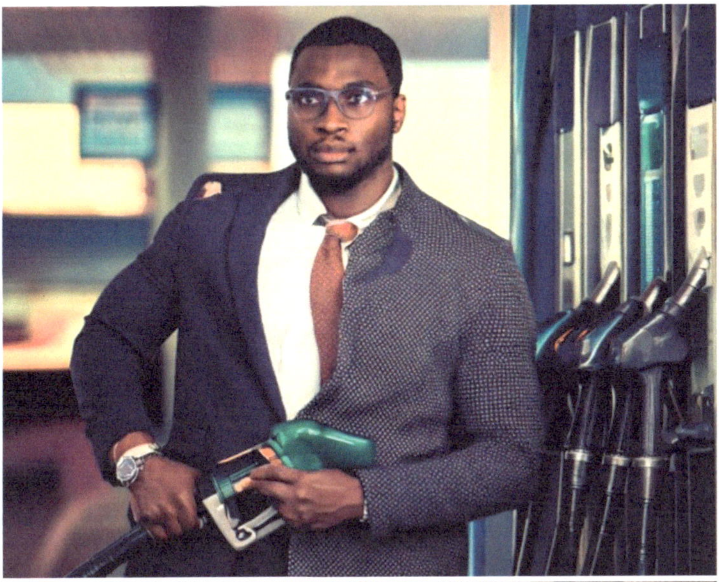

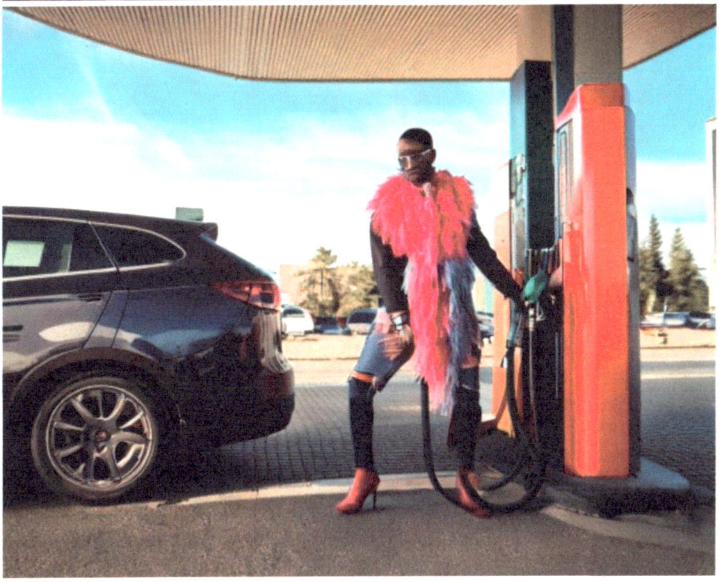

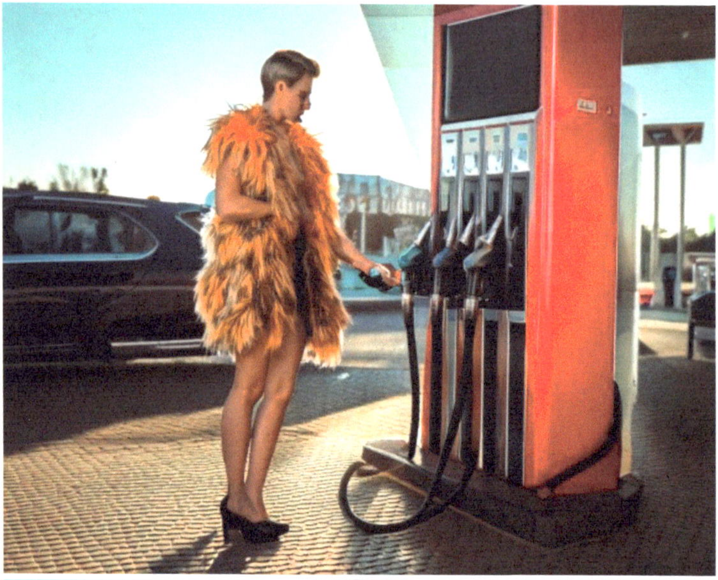

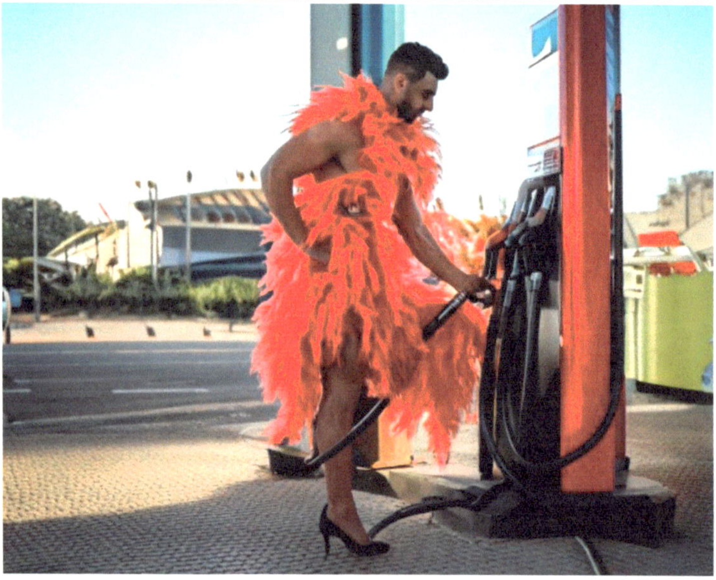

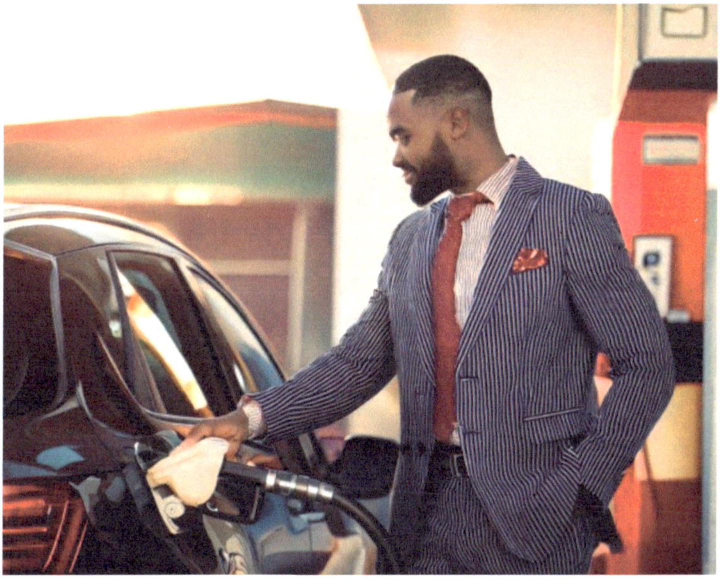

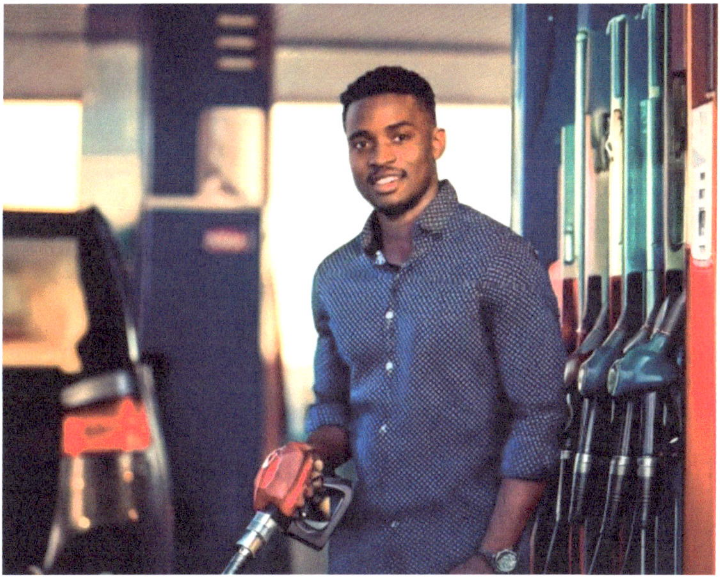

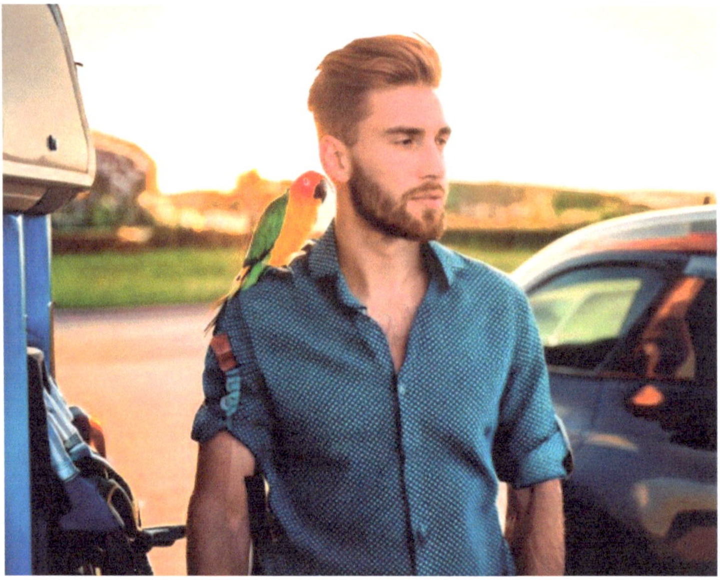

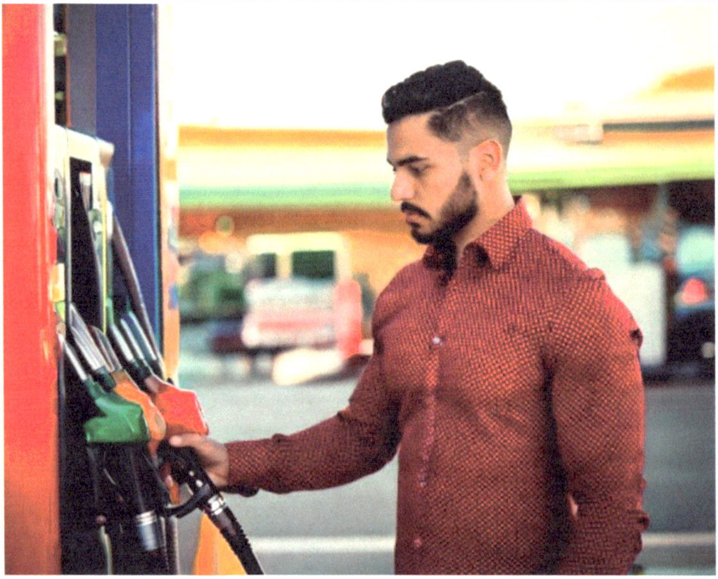

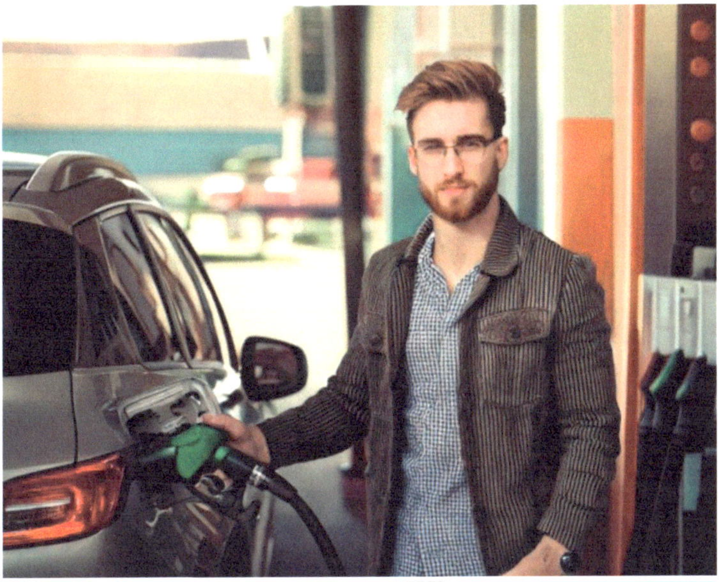

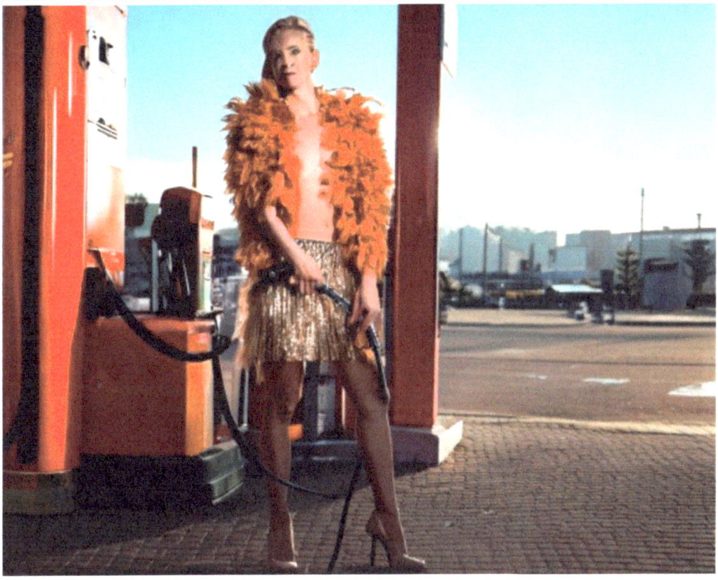

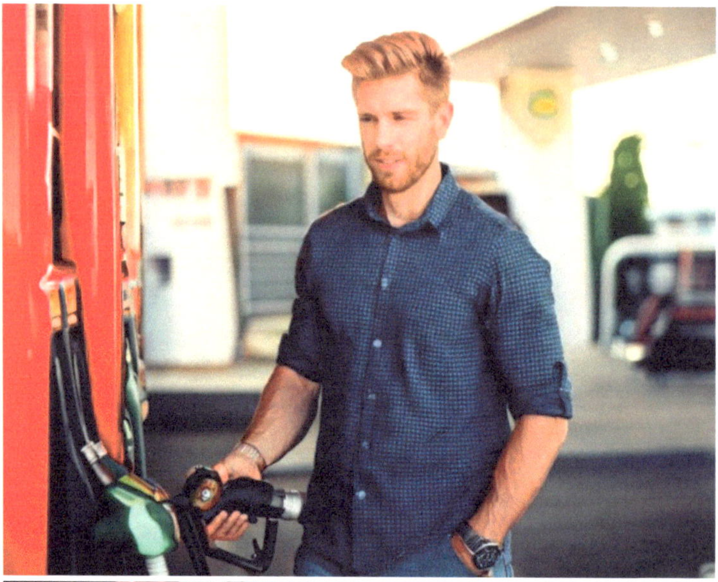

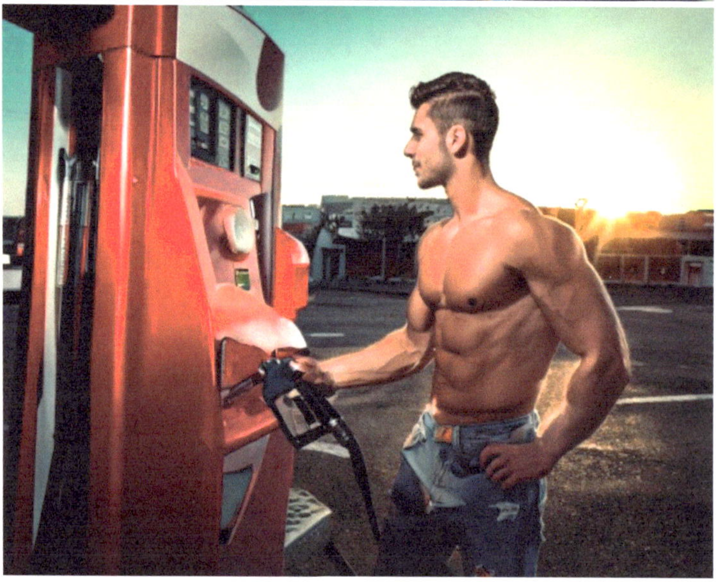

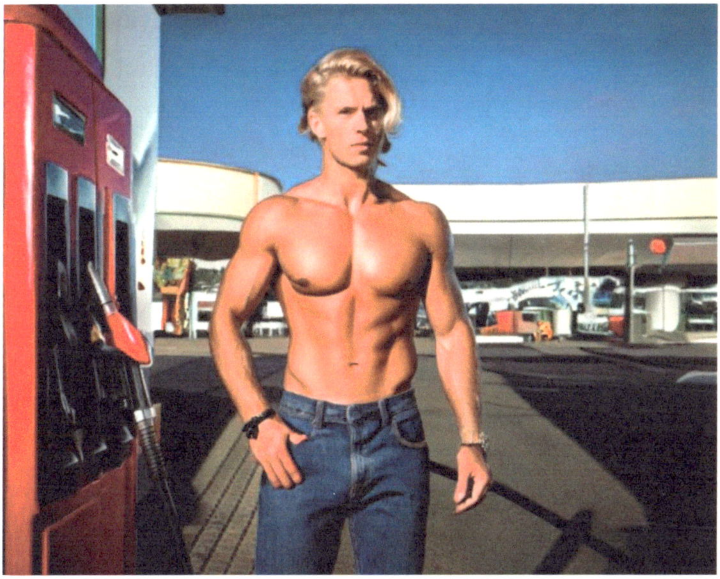

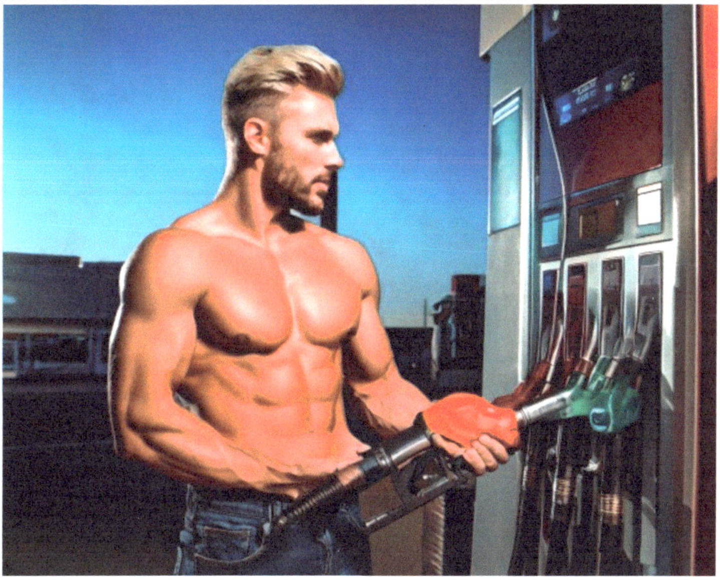

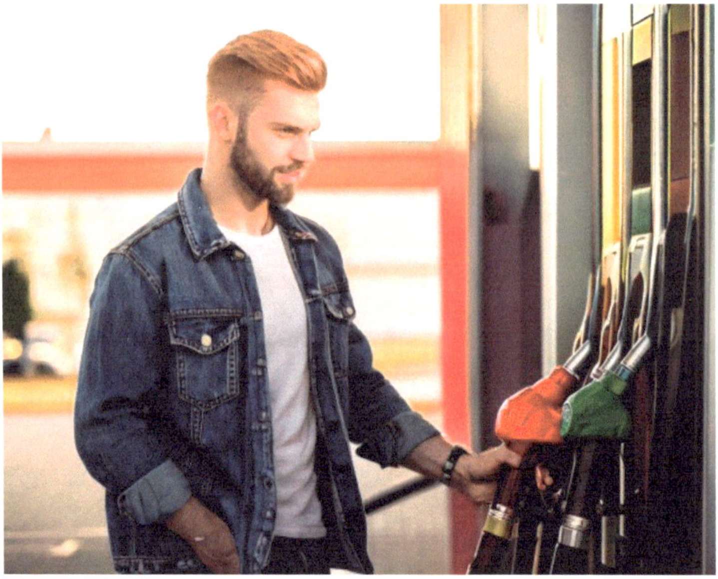

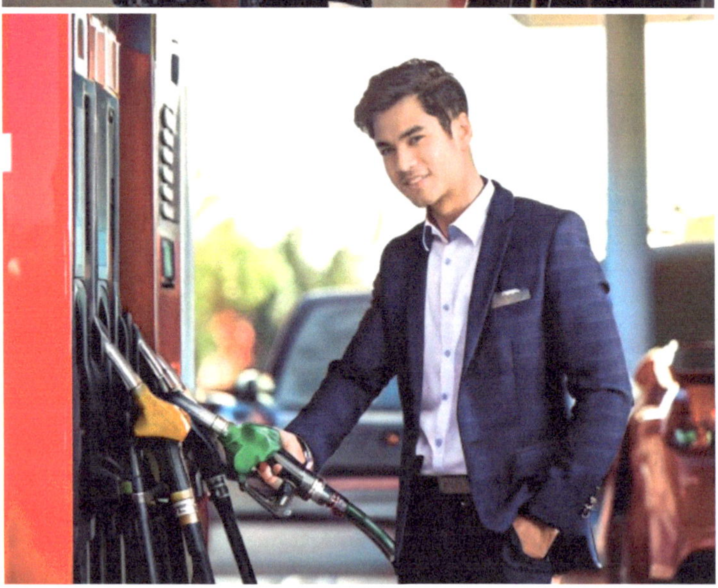

Excerpts from found poety by qz blaze

PRACTICAL
MYSTIC (91)

PRAIRIE FIRE (94)

LIT

Almost 94

Glasnost

Mood Changer 92

Popping poetry

Like it was

Pills

PRAIRIE
JOURNAL (92)

BURNING

The bullet stuck under his
dick. 92

He lived flaccid

PRAIRIE (94)

BALLADS

I'm so tired of this flatness!!!!!!!! 92

And poems that make fun of fatness

PRAIRIE
SCHOONER (94)

SAILING ALONG

I'm tired of people dying!!!!!!!!!!! 91

What do we have here? --

Words to remember

People

Or

People

To remember words?

Say Fou! 94

For form or our four

Or four for u four

Or for our form

Or four for form (Brava, Eleanor)

Permafrost 93

Froze itself out

How Can She
Come? 92

Unsheathing and masturbation

IN THE

PRAIRIE
WIND (94)

PRAKALPANA -- ROBISENA
(94)

Pointilism poetry 94

Without the ineluctable

Point

Poison pellet penned on 93

Poison producing paper

Dioxin to describe

Cancellation of chlordane

Inging 91

All

Fukka

Up…..uh…upward

A sideways rhyme (not quite absorbed)

Back and forth in time (adsorbed, maybe)

End rhyme on the last word

Not quite absurd (parenthesis for attached entities)

Don’t eat meat 93

Don’t eat anything

Eating kills you

Everything is everything 92

PRESBYTERIAN (
94)

PRESCOTT
STREET (92)

If I wanted to cater 92

To the highest common

Denominator of humanity --

What would that be?

PRESS OF
MACDONALD (94)

PRESS OF THE NIGHT
OWL (94)

PRESS OF THE THIRD MIND,
BILE (94)

One mind to mind my own
business 92

One mind to mind reading this poetry

And then, I have a mind to throw it away

He carries his walking stick like a
bow 94

A Native American in the shadows

Concrete under his feet

Over his liver --

no

Over his spleen

no

Over his pancreas.

On the return trip

He balances a comfy chair

On his head.

Hey, hand me a
twenty, 93

I’ll give you a wink.

For thirty, I’ll wink

At you …. in Spanish.

STYLE

PRIMAL VOICES VISIONS AND REVISIONS (93)

PRIMAVERA (94)

Right Rite Wrong
Wrong 91

Left loved or loved left?

Wronged love or loved wrong?

Left right or left wrong?

Bubble 94

ubble

Bub

bub

ub

Slippery 91

Full-flavored fish on the tongue

We don't all reach adulthood

United --some of us never do

Some of us stay separate.

PRIME TIMES (91)

Hard to Make You
Count 91

A list of lists

Or a list of live things

Is as hard to count

As a list of dead mementos

So I add a dropped angel wing,

Ripped from her left side

For your violent poetry

Your "Mommy! Mommy!

It's a mummy!" lines

Your obituary imagery

Your sea-shore sloth

Preferring poetry penned for lazy tongues

Like aural sex without foreplay for play

Depending, so much, upon who does the reading

And for whom

PRINCETON (94)

On chalky lines 94

Of stadium imagery

And contrived crowds

He plays an away game

Of the poetic mind

Contradiction 92

Too wide to admit

Darkness

He's an economy-sized

Asshole

Life 93

Is a missed

Tree

PRISM (94)

Camel's squished through eye 94

PRISONERS OF THE NIGHT AMARANTH (94)

Weep not for Argentina 91

Angels' echoes 92

Angels, angels

Flock of seagulls

Descending angels

Down to the depths

Of my crazy, little, manic

Up/down ma, man, manic, maniac, maniacal,

"N"-"U"-"T" spells "Nuts",

Poetic heart.

Hey, Masochist! I'll give you a sharp kiss 93

Apocalyptic Entrails 94

Cryptic conclusions

Tryptic-folded illusions

Chipped dick delusions

End-trail conclusions

A lah lah lah 93

A dee dee dee

I want to be

Frust-

rated

AS GOOD AS ANY,

I'M TAKING THIS EXCUSE

TO BE TOTALLY FREAKED OUT

PROEM (91)

THROW `EM INTO A BLENDER TO MEND HER LINES

PROOF
ROCK (94)

BEDROCK BARNEY RUBBLE ROCK
HEAD/ROCK-RUBBLE

Reading--can it be
taught? 92

PROPER TALES MONDO HUNKAMOOGA
DWARF PUPPET COPS GOING FOR DONUTS

(91)

This poem! 91

It actually sucks --

But not in (all) the right places

Ante/Post 91

Were other people here too?

PROPHETIC
VOICES (94)

Poetry is n

Now,

We are fat-free poly-unsaturated

PROTEA (93)

Over grass meadow, 93

Mirrored birds,

Breathless,

She went

Barefoot to the Buffet

That smell was
deodorant 91

Pheromone

And the genius of the balls

PROUT (91)

GROUT,

ROUT IT

OUT

IDEAS

Our protozoa 91

Dance together

Circuitously

Please stop this!!!!!!!!!

PSYCH IT OUT!!!!!!!!!!

THIS IDEA ISN'T WORKING!!!!!

OUT! OUT!

SO FAR FROM

PROVINCETOWN
(94)

With not always tongue in
cheek 94

But, at least, tongue in twat poetry

Living In Plasterland 92

Barbara sees

"Twilight polka dots,"

Vases n' Pottery

Shards of clay

Lying broken on broken shelves

Like living in an archaeological dig

An anthropological ghost town of an industrial area.

Pimp stays a respectful distance away

From his hooker.

This uniform is the only thing which distinguishes me

From the girl beyond the fence --

And the fact that she's her own boss.

Her pimp is like a therapist.

I listen to them fighting,

Late at night, after business hours.

She's the boss in that partnership.

She's the boss, all right;

She brings home the pork.

We avoid each others' eyes, across the fence,

Each thinking the other criminal

Or at least negligent of some thing

The other does without, but we could never.

We hardly ever see those shadows

Throwing dead bodies into the Dempsey Dumpster;

The smells are the first signs of invasion, this Summer.

The insects conduct the funerals for those

Smells. The pimp sees my feathered ear-rings

And calls me a dinosaur, winking, at a distance;

Illogical leftovers of the sixties

And the only logic for the "homeless"

Squatting on toilets with no running water,

Walking the streets at night,

Patching up knife wounds with clear plastic packing tape,

Is not to read the polka dot twilight

As deep or brooding

“`Cus that would be a bad trip, man!”

And not to lay our bodies down for too long.

I sit against the building I’m guarding, as day fades,

And wonder

“How long will it take for the fire ants to find me?”

PSYCH IT (94)

God only knows where I would
be 93

If not with him, all those years, abusing me,

But there’s no time to look back there

And only bad, bad, bad, black memories

Can you smell my
anger 91

At having inhaled

The raw, rotten eggs

Of your arhythmic verse?

PSYCHOPOETICA
(94)

When
advertisements 92

And Hallmark

And the war effort

Take the pretty and heroic poetry (for
effect)

Then poetry must be

About something other than (just) flowers

And sunsets

If cut flowers and the sunset

Of life, the inevitable death

Are okay (for adverts, Hallmark and
propaganda)

Then poetry must

"Rage, rage against" it.(them)

PTERODACTYL (9
1)

DEAD

PTOLEMY (94)

PUCKERBRUSH (
94)

Ultra-radical 94

Big, bad bully,

Littering

Little pay-checks

Along the roadside,

Free radioactive pennies

For children to pick up

All these poets trying to be remembered in words 91

All of us, poets,

Trying to be remembered

In words

Rather than words

Remembered

………………..

…..........BUT

"I did not kill no cat" 94

PUDDINGHOUSE (94)

A PUDDING PUDDLE

Blister-wrist 92

Time takes care of the rest

Verbal directions 94

Like a director talking

About the scenery and lights

Blueprints in words for (/an invisible canvas in the

minds (of artists who are tired of/ painting by numbers,/

These are not all that poetry is.

Artists to paint in their minds

Are not all that poetry is.

PUEBLO (94)

PUERTO DEL
SOL (94)

PERDUE (94)

PURPLE
PATCH (94)

Puce Debussy 94

Cyberspace waves

Pusillanimous tones

The terror of error

New symbols to transpose into

Art/artisan/work translations

And the mother (La M

Astride the information highway

She shrinks from the sun

Never sees their son

Weighs three quarters of a ton,

But she's in there, on the bus,

Ms. missing, Ms. mini-millionaire, Ms.

Autoexec

All writers are 92

Great

Graded

Grated

rated

But not all

rate

And not all

ate

at

The celebrity plate

Memories of Max? 91

Thrill me to the min --

You know, that's just a few blocks away

From farago

Where he flashed, reflected

In your polished, silver bowl

Like a ladle

PYGMY
FOREST (94)

PYRAMID (93)

QUARRY (94)

I met a maladroit? 92

Do they really say that?

Really, do they?

QUARTERLY (94
)

QUARTERLY (94
)

QUARTOS (94)

FOR THE

QUEEN OF ALL
HEARTS (94)

Rodin was not
automatic 92

Rilke was something else

AND

ELLERY
QUEEN'S (94)

15 inch knife 94

Or standard 6 inch blade?

Always propaganda?

No truth, no news to use,?

Always the half-back theory?

Always the O.J. monopoly (FLA, CAL-TEX)?

Always some football legend

Caught up in the black myth?

QUEEN'S
QUARTERLY (94)

BOUT BETWEEN

QUICK BROWN FOX AND MAD
DOG (94)

Soldiers! 91

You went through Soldier Hell

Don't tell the story

THEY want you to tell

The story you think we thing we want to hear

RACKHAM (94)

ENDS LIKE HAM

ON THE TORTURE RACK

I GET MY INSIDES SMOKED

Puerto Rico is a puppy? 94

I didn't say that

RADCLIFFE (94)

OVER A RAD OCEAN OF STARVING PEOPLE

The 92

Blind

Can

Describe

UNDER

RADDLE
MOON (94)

RADIANCE (94)

So heavy 91

So heavy, you hope

It's all fat in there

A collector's journal? 92

Fat collector's propaganda?

Do they have diet bubble-gum cards?

Liposuction leftovers in sealed jars?

This head is a big head 94

nuff said

RAG MAG BLACK
HAT (94)

HAIR -- FORMAL, THAT'S ALL

Do you slaughter your brain 91

Like your cow?

Is that simplicity?

Is that my Elsie?

The fan on the IBM XT 93

Winding down after the circuit's cut --

That's an eighties metaphor for death

During a blackout

Boy, oh boy, there's a dead body,
there, 94

And this is a poem about it

RAINBOW CITY
EXPRESS (94)

YOURSELF ON EAST 105th STREET

Art 92

"Unseen" except in poetry?

How does one unsee it?

WHEN

RAMBUNCTIOUS (9
4)

Ooooh, I wish! 91

I'm round 92

But too fat

To bounce

RANGER LION
ROAR (94)

ESCAPADES

A ranger seems 92

A noble thing

Until he creams

Over my girl, Pearl,

And has his friends

Break my left wing to keep me away

From hurting him

And though I'm lesbian, mostly now,

This is an eagle song

FOR

RANGER RICK (94)

TO WRANGLE

RARACH (94)

RARITAN (94)

RASHI (94)

RAVEN (94)

PICKS AT

Vulture 94

Hidden

In the raven's shadow

RAW DOG (94)

Welkmance 92

Champagne bubbles

And nothing on t.v.

TITLE 94

Sometimes

The whole meaning

Is in there

READ ME

REAL (94)

I'm so glad 91

I'm so glad

That you are alive

On the Gonads of Grief 94

We are worms in a can

BUT NOT TOO CLOSE TO

REALITY
STREET (94)

RECONSTRUCTIONIST
(94)

FORGETFULNESS

Come share my
yearning 94

For… well, not hurricanes nor watermelon

RECORDING ANTERIOR
BITEWING (93)

PHILOSOPHY AS

Lens Occlusion 93

Hunter with scope

Thinks he's bigger

I admit that I am an asshole 92

But I don’t dwell on the thought,

Just write poetry about it

“THEIR PROBLEMS” IN THERAPY, NOW

RECOVERING
BLACKBERRY (94)

STAINS

We tear at the scabs 92

Weave word butterfly

Bandages

To seal the wounds

Frigid dark 91

Frigid dark

Frigid dark

OF THE

RECURSIVE
ANGEL (93)

Future Poet 93

These lines

Are not enough

RED ADLER
PAN-EROTIC (92)

WASH-OUT-MY-BRAIN-WITH-SOAP SOMA

Fish Flowers 91

Spurting Flowers?

Around the world?

A space shot?

We know so many names

For flowers

Like Eskimoes have all those names for snow

How many names do we have for love?

COLORS AND FLOWERS AND DEATH

AND GOSSAMER

VEILED SEX

EROTICA, BUT NO PORNOGRAPHY

REVERSE DISCRIMINATION

REVERSE HYPOPARISTALSIS

REGULAR OLD LIFE

HEIGHTENED TO THE HEAVENS

OF UNREALITY

BOREDOM AND GRIEF

CHEWED LIKE DELICACIES

IS THIS ALL THERE IS FOR POETRY?

NO MORE SPINE TINGLES?

NO MORE LIFE MADE FRESH?

MEMORIES, MEMORIES

OF OLD IMAGES, IN CREATIVE COMBINATIONS

BUT NOT A NEW VISION

OR A

RED BASS? (91)

They dump their waste on
us 91

Saying, “you want the good life?”

Hide maps of contaminated areas

Like they hide their combination safes

Inland revenue (for the children)

Pay tax

To play clean

They dump their waste on us

Saying, “The reason it smells funny, like
that,

Is because it’s art”

RED CANDLE
CANDELABRUM, (94)

A GOLDEN THOUGHT

The nurses talk about
snakes 91

Remembering the giant ones

And the little ones

And those that were very different

Maybe the brightly-colored ones

Or those that were some sort of first

But don't remember the last

One they see before they're bitten?

Tomorrow, dreams
begin 94

No

Today, dreams begin

No

Yesterday, dreams began

No

Yesterday, dreams had been beginning

No

Day-before-yesterday, dreams had begun

RED CEDAR (94)

SMELLS

Dr. Spock's Little Monsters 91

Some, twisted, turn

On the memorized road

Of "for a nickle I will

Blow up the toll gate."

Christmas Memory 94

Didn't need to go to Times Square

To see that

I REMEMBER THE

RED
DANCEFLOOR (94)

Confetti silk 92

Incipient stone

Cat-fight clammor

We're both in love

With the same woman,

Inside us

Pinched nose 94

Between his crack

Two birds with one

Asshole

RED HERRING
MATRIX (94)

OF

RED RAMPIN' (94)

REDNECK (94)

NOTIONS THAT WE MUST SMILE, BUT

When women go
mad 94

In Texas

They stack the men

Behind the dryer

REFLECT (94)

EUPHONIES IN ACOUSTICAL MIRRORS

REFLECT, YOU PHONEYS

REFLECT ON THIS --

Why ulcerous oysters? 92

Why the false consonance of psychotic passion?

Why the vowel play ploy of sterile rigid?

Okay, Faulkner's been a bit neglected

Vermillion vermis worm

Mocking the jock? A white orchid oyster gypsy

With tan feet? Ooooooooh-- she and she.

Winter is a Korg
sound-module, 94

Empty until you put some tones in

REFLECTIONS (92)

CAPTURED IN PRIVATE MIRRORS

Editing of 14 year old's 91

Words

Makes she (her?) look like

She wants her too

Dissappointed tears -- 92

Should have talked to the receptionist

To make an appointment

REFLECTIONS (91)

WHICH ARE CAUGHT MOMENTS

OF PHOTOGRAPHIC REALITY

The angry poets 91

Who feel and know great injustice

Only to be silenced --

Are they included in that

"All who love?"

Then, who is not a poet?

ARE

RENDITIONS (94)

OF UNREAL EXPRESSIONS

The rocks, pressed by
waves, 92

Are not silent --

Only, sometimes, unheard

EVEN

RENEGADE (94)

FIGUREHEADS

Witchcraft 92

Is the way

Of the ultimate follower,

The gladly taken away,

Not the renegade

THE GUIDING LIGHT OF THE

RENOVATED LIGHTHOUSE (91)

Images in the flashing light 91

To avoid what things mean to me.

Is it enough? Is it enough

To let the heart know it means… something?

Isn't that something like a moon without a sun?

Like an asteroid in a black hole?

REPOSITORY (94)

RESIN (94)

THEIR FACE

Knee Legacy 93

Spiked shoes on astroturf

RESIGN IN

RESPONSE (94)

So, Rilke lives? 92

Like Elvis?

RESURGENS (92)

More Babel-babble 92

REVIEW (94)

REVIEW LA
BOOCHE (92)

THE BOOTY, THE BEAUTY

Love is a poet unidentified 92

In the right setting --

Odds 7:5 (demographically)

REVIEW LATIN (94)

I'm beginning to feel like a
Youth-Vampire 94

Blow your breath cold

Rebreathing

Life into me

REVISTA (94)

RFD (94)

I am a light among lights 92

A spark screaming, inside brilliance,

Not "and nothing I remember

But the forgotten" --see how it works?

IN

RHINO (94)

"My-yyy Grandma was 94

Stupider than

Youuuuur Grandma!"

I would like to say 93

This poetry

"Shimmers like a July rainbow"

But this is not true

VERSIONS OF

RHODODENDRON (
92)

REMEMBERANCES

Body-fluid poetry 92

Spurts

Into empty space

BUT NO SMILES

RIALTO (94)

Zen 91

Suicide?

Rowing into Summer
darkness 94

In only briefs,

He calls himself, “Poet”

THE

RIDGE (94)

OF MY FACE

I put a flower on the bottom 92

Of these lines so that someone

Will think they might be poetry

Iris

RIENDA SUELTA, GIVEN FREE
REIN (93)

RIO GRANDE SE LA
VIE (94)

SLAV VEE

SLAVERY

TO THE RAIN

Rushing River 92

It's not fair to the delta

To be so close to the ocean

With you,

Seedless, tired, running away.

RIVELIN
GRAPHEME (93)

NO! COMPUTER GRAPHICS DO NOT MAKE UP FOR THE SMILES LOST TO

RIVER CITY (93)

RIVER RAT (94)

Perhaps, it's a ball game 94

The sun is out

He scratches balls, rolls bat in fingers

Sheet-Metal Husband 93

Sheet-metal machine

Sets the rhythm

For our sex

RIVER STYX (94)

A sound! 92

A sound from the faucet

Flowing into the sewer, the bayou

Then the ship channel

Then the bay and ocean

A sound, a pattering of rain.

Zindi jahh janily rindow 94

Iddle ahroock tanzilly love

Rashandunah guberick dewerdin

Eddmindoon finnelle galuzin dranook

RIVERRUN (94)

LIVING

Oh, Artist In-vide-ous 92

in

video

us

Is easy poetry better? --

Light beer, light beef

Light poetry, or is it

Invideous?

Deer-Poet 94

Runs forest paths

Scavenging for death

ON THE

RIVERSIDE (94)

Skull skull skull skull skull 94

I think I'll call this

River-Rhythm Poetry

(With analogous oars)

OF MEMORY'S

RIVERWIND (94)

Sometimes I stand on my bed 92

And try to piss into the lake

Does that make sleepy sense

To you, watching me,

My belly protruding,

Hips tilted like a man

THERE IS A

ROAD/HOUSE (91)

Up your ass, 91

You flaccid pig!

They should call you

Todd Lesse

FOR NEURON

ROADS (92)

Before making love 92

Lying minds

TO

ROANOKE (94)

I’m sorry but the words 92

Aren’t inspiring me

So, I will use less of them

OR

ROCK FALLS (94)

DOWN DOWN

Butte-first 94

Echoes

INTO

ROCKFORD
TRIBUTARY (94)

ECHOES

Does grammar matter? 91

Does Grandma?

Does new poetry disconnect us,

Looking for obscure connectives?

Writing 92

Lines in car

Auto accident

Tonight, I will not talk 93

About death at all --

Oops! -- Already did it.

Diamonds and death 94

Flowers and death

Cold wind and death

Dancing and death

Cars and death

Poetry and death

Death and death

BUT NO SMILES FOR

ROCKY MOUNTAIN POET
MONTANA (91)

Don't juxtapose your words 91

If you don't want your Cowboy

Doing it the way he did howl

"Up giddy! Up giddy!"

ROCKY
MOUNTAIN (94)

TETON/TITS

Was Cezanne a peeping
tom, 92

Highlighting bather's breasts?

Was he not a man?

ROHWEDDER (94
)

IN THE SWEAT OF CO-EDUCATIONAL COMPETITION SKULLING

Dream shredder 94

Talking Tolstoy to justify

Common Bovary's

Stretching Zola stocking-stalkings

Sometimes, I pass 92

Words

In thick chunks

CIA, dropping 91

Palm tree microphones

On the Ho Chi Min Trail

ROLLING (94)

IN THE DOUGH

Every martyr is sacred 94

Every suicide is sacred

Every life is sacred

But some tongue-less people

Work on computers

From the “girth of expulsion” 91

I go “pop”

Bone dirt 93

Gas 92

From dead body

Yellow cardboard

That plastic smell

Of old photos

ROMANCING THE
PAST (92)

Tarnished with newness 92

ROMANTICIST
ATLANTEAN (92)

SNUBS

THE
ROMANTIST (94)

Mass graves to hide
swollen 92

Reasons

Sand turned to help blood seep

Away

We go on with a certain

Justice

But justice

Is

Just

Ice

In this desert

BECAUSE THERE'S NO

ROOM (91)

NO

ROOM OF ONE'S
OWN (94)

ONLY

ROPE BURNS (94)

OF CONFINEMENT

Christmas cowboy 94

Stuck by his spurs

In the chimney

ROSE SHELL (94)

CONVENTIONS

Automotive Mother
Goose 94

Fairy tales for women

And a few sensitive men

(The rest of the sensitive men

Can run home to mama

Like the cry-babies they are)

THAT DATE BACK TO

ROUND
TABLE (94)

DAYS

Dream weavers 92

Should be

Dream plasterers

Plastering the cracks

ROWANTREE (91)

RUBBER
PUPPY (92)

BUBBLE PUPPY

FLOATING ABOVE IT ALL

Bagel Dog 92

Has tire marks

Where the car ran over

His back legs.

He'll never but crawl

Maybe we can build him a dolly

To skate over the hyperbole

Of "days that tread on" your "face".

THE

RUGGING
ROOM (94)

ROLLING OVER

RUNAWAY
SPOON (94)

S T I C A L I C T H S S T O O P I
D 91

The barn 93

Swallows

Wind

Singing the greens 94

Because 8 percent of males

Are color-impaired

RURAL
HERITAGE (94)

I'm almost used to not knowing where I'm going 92

Since they bought that trailer and sawed off my tail

I get the eerie feeling they're going to make a show

Of me. (At least give me a luxury ride to the glue

factory)

THAT

SACHEM (94)

You offend the way 92

Vomit offends the willow

These bubbles glistening
down 94

My bosom are my toys, my treasure,

My perimeter. These baubles glittering

On my bosom are .....

Men want them

Removed before they see me

CALLED

SACKBUT (92)

Lick my coins? 92

SACRED RIVER BAY AREA
WOMANS' (94)

SACRIFICE THE COMMON
SENSE (94)

Turtle crossed the road 93

And I know why

Pain, pain 94

What are you telling me?

Now's the time

To write poetry?

SA-DE (92)

SADISM OF SACRIFICIAL ALTRUISM FOR SHOW IN

Seasons of Contempt 92

These new laws like

"The Freedom of Access

To Bathroom Entrances Act"

These new laws are deadly

Are killing my business

Retort!

Re-

Tort

Retort, but all there is

Is retarded reaction,

Reticence to move the legal

Mountain for this molecular

Tart who screams out into

Blank lust, “I would fuck

The whole world for freedom!”

This tart tastes the tramp

Trickle on my tongue

I can’t blow the jobs back

In this foul, flatulent wind

Of world order wishfulness

I bow my head, beaten

But will bounce back……

Now, I’m on the attack!

ST. ANDREW (94)

At this point, 92

History has a right to say

Of me and my sisters

That everything bad

Said about us was true

That we have been stupid

And whining, trite with

Hacked-knees

SAINT
ANDREWS (91)

ST. ANTHONY
MESSENGER (94)

Charity begins 93

....Who cares where?

Saffron 91

In rice

Clumps

To clamor over

Hubby-bubby 94

You're a froggie!

Happy anniversary

ST. JOSEPH MESSENGER AND ADVOCATE OF THE BLIND (94)

Snow-blind, 92

I need shadows

To see

ST. LUKE'S IRIS (91)

PETAL WORDS AND

Those snakes from 91

Hoop-snake cartoons

And Indiana Jones

Inspired this poem

ST. MARTIN'S (94)

SALMAGUNDI (94)

The Skidmore girls 92

Sneak dates with

Poor poets who have

Earthy, real rhythm

But publish Seamus

In wedding invitations

`Cus he's …. o-kay

SALMON (92)

AMBIGUITY

What there was of daybreak 92

Dribbled

Over skyscrapers

SALMON RUN (94)

LIKE STALKING MEN

Catching running vines 94

Of dewberries for breakfast

Took until afternoon

SALT LICK SAMPLERS LUCKY HEART (94)

BELIEVERS

Babba-loo-ey 92

Baloney

In a full-length

Mirror

IN

SALTHOUSE (91)

What there was of
daybreak 91

Shouted,

Muffled in the thorns

Of our wedding-bed

SAMISDAT (92)

Empty frame 91

Where our first dollar

Should have been

Death is not the poet's 92

Only job.

My favorite poets use it

Like a garnish

SAN DIEGO (94)

Prego, prego 94

Too much salt in the waffles

And I thought the best

Ingredient, the big-name-chef-

Ingredient, was supposed to be love

Orchard Memories 91

Trees growing back

But what does not?

What is this sense of loss

I feel?

What have I forgotten, here?

We were never not lost 92

On this backslide of a booming

Generation,

We were never not lost

SAN FERNANDO CERULEAN
MINI (94)

MIRACLES

Okay, if I agree 94

That there's no use

That all is lost,

Will you still strive

To be a cunning linguist?

Will you test your talent on me?

Can't get ahead 91

Can't get head

Can't get my head

On crooked enough

To make some real money

Like they say, 92

Bullshit walks,

But they don't mean

The kind that just sits there,

Steaming

SAN JOSE (94)

Mountain woman, will you stay there 94

Stuck

Where you belong

If I write a poem or sing you a song?

Sissy-pus 91

Depending upon your friends

To "hold me back!"

SAN MIGUEL (94)

Charles should write as 94

Charles more slowly

Squeeze his effusions

Into a tight stream

SANDBERRY (94)

Waking up with "blue sight" 93

Your blue flowers

And blue sky

Would look

Either black or white

Hey, poet, hey hey 94

Help me make it through

The day

DINGLEBERRY FOR A

SANDPIPER (94)

SCRIBING

SANDSCRIPT CAPE COD
PROXADE (91)

PRELIM

My lines are only good 92

When people watch me dance

SANSKRIT (94)

SCRATCHES

Hey, Cristy, I'll put my tongue around
you 94

You don't mind fresh fish, do you?

I'll call you flounder-face

And you can keep the name as your own

Poet reads the obituaries 93

To think up new ways

To write about death

IN THE SAND OF

SANTA
MONICA (94)

AND DOWN THE BRONZED BACK OF

SANTA
SUZANA (94)

To describe yesterday's
Promise for the future
Looking-at-a-painting- 92
That-you-can't-see-
But-I-don't-give-a-fuck-
Because-I-can-poetry
To dam or not to dam 93
That is the fishy question
ALTOGETHER SCREAMING "IT'S FLOUNDER , IT'S STRIPER!"
SCARP (94)
As a poetess 94
And a poet
I have double indemnity
Barney's, New York -- 92
That's where daddy
Got his suits.

So, what?

So, they were good suits

OLD POETS KNOW ENOUGH TO PRETEND TO PAY ATTENTION TO HER

OPINIONS UNTIL THE SAND FLEA SAYS

SCAT! (94)

The Altar 94

I found the title

And expect you

To do the rest

Of the work of

Knowing what

This poem is

WHEN THERE'S NOTHING THAT BINDS LIKE PIPE STRAP

SCAVENGERS (94
)

SCATTER

Garp on, dude! 94

Armstrong's 91
Poetic footprint
Left
(Or was it right?)

Poet lost 92
In cyberspace --
Can't make a game
Of words
With only one winner

Loaded pistol 93
That's all
It just sits there

(EXCEPT A KILLER FROG
FROM
SCIENCE FICTION
STAR (94)

My poetry has 94

A footprint fetish

"Hey, Stewy, this is Fred 92

Just thought I'd call to make sure

You knew there's a crumpled body

Of dead poetry on your front lawn

And it's starting to stink up the neighborhood"

Silver million brain drip 93

Moons so wave spatter sanity

Melting good-bye down lunar sky

Madness becomes a voyaging eye

High and up in the Earth-bound moon-ship

A your, your, YOUR

IN SCI-FI HANDCUFFS) CUS THE MANTA RAY

SCOP (94)

BITTER, READY TO SETTLE SOME

SCORE (94)

AFRAID OF AIDS BECAUSE OF THAT

SCRATCH (94)

FROM A JUNKIE'S NEEDLE

Poet, But Not For Long 94

I came to the conclusion

That anything human

Is too tainted to touch

When 3: PM dies 93

Ian cries

AFRAID CUS THE SEALS HAVE IT,

`CUS DOLPHINS HAVE SOMETHING,

AFRAID OF THE

SCREAM, ALTERNATING
CRIMES (92)

IN PERPETUAL SURPRISE
CONFIGURATIONS IGNORE THE

Redundant ambiguity 92

SCREAM OF THE BUDDHA AFTER THE
BUDDHA ROSE IN CRYPTIC WHITE (93)

Casual, anonymous
suicide 93

AMONGST

SCREAMING
TREES (92)

SCRIPSI (92)

He may be a poetry
reviewer 92

But is he a reviewer reviewer

Reviewer reviewer reviewer?

NO RETORT HERE

SCRIPTOR (92)

Poetry For No Cause 92

From an immortal Autumn

And select songs of scarlet

Procession, in perfumed evangel passion

Of tumbling crystal and windflight,

Swift moon, slipping into ardent freefall

Become elements and, with their velvet vessels,

Press to swiftly multiply

SCRIVENER (92)

Slap that bat's ass 92

If it eats your wristwatch

Leave it alone,

If it only wants insects

Such landscape 91

Such sun

Such an open door

But, where's the

"Girl in the white wetsuit?"

SEA FOG HARMONY VOICES INJUSTICE FOR A JUST FUTURE (91)

But do you love those old men in the park? 91

Or does it take a real

God to do that?

SEA TAILS (91)

The sharks ate the mermaids 91

Now, you can fear the sharks

FOR A

SEANCE (93)

IN

SEATTLE (94)

She ladys her headlights into me like 94

Bumper car cushions, her electrical switch expertise,

The way she fingers it should convince me

She's got a man

Wants the switch deep inside

Guided only by Chilton's

Her book of turn-ons and turn-offs

UNANNOUNCED UNTIL THE

SECOND
AEON (94)

BECAUSE THE GOVERNMENT KEEPS
EVERYTHING

SILENT, NOW, EVEN

SECORD (BIG
TWO-HEARTED) (91)

There IS immediate
danger 91

Retort, retort, retort

EVEN

SECRET
GOLDFISH (92)

Icarus Hiccups 92

Dad's hot air

And sporatic spasms

Are killing me

SECRETS FROM THE ORANGE COUCH (94)

Movement between tension and release -- 94

Music, like that, encounters Xeno's paradox

Cramped smile 92

Is like a

Cramped simile

OR SO IT

SEEMS (93)

Black dreams 91

Talent eyes

Taloned eyes

"Got no sight" poetry

SEGUE ROOF
SEGUE (94)

ROOF LEAK OVER

SENECA (94)

Let the poem do
something! 91

PATCH A HOLE IN THE

SENIOR EDITION COLORADO OLD
TIMES (93)

OF DROWSY

SENSATIONS (93)

I want my angel,
yesterday? 93

Cirrhosis P.C. 91

Sexist, Papa-Basher!

Mama sucked `em down

As good as any Daddy did

AFTER ALL THE

SEQUOIA (94)

ARE CUT

Is it sexist to say most rapists are
guys? 92

Would it be sexist if I were a celebrity?

THE SERPENT AND
EAGLE (94)

EXTINGUISHED AND EXTINCT,

SEVEN BUFFALOES PLAY BLACKJACK WITH
THE GRAPEVINE AND HOLLER (94)

Paint pretty pictures in my brain 94

And then snap small bones

In the “snowy field of foxtracks”

Dripping blood fluids, all about

THAT THERE SHOULD BE

SEVENTEEN (94)

OF THEM

NORTHWEST OF THE

SEWANEE (94)

Words For A Poet's Funeral 92

"That elegy really sucked"

AND NO FORCED

SHAKESPEARE (91)

Corporate power 91

Is the perfect tyranny

With no succession

To kinder kin

And no queen, there,

To let love sink in

BUT, BUT, AT LEAST........

THE BEAUTY OF SHAKESPEARE

OVERCAME BODY SMELLS, STILL DOES.

TODAY'S POETRY IS SO FRAGILE THAT

IT CAN NOT STAND TO BE PISSED ON.

TODAY'S DELICATE POETRY

DOESN'T DARE EVEN TWEAK

THE NOSES OF UNDERGROUND MONARCHY --

I REBEL MORE WHEN I STRIP OFF MY CLOTHES

THAN DO THE IMAGES OF PASSIVE-PAINT-POETRY.

USUALLY A WAITRESS, I HAVE TAKEN OFF MY TOP

FOR MONEY AND FAME AND IT MAKES ME MAD THAT

IF YOU ARE AN OBNOXIOUS,ANTISOCIAL,

UNCOMPROMISINGLY SELFISH,

IMPOSSIBLE-TO-LIVE-WITH-BITCH

WHO GETS PREGNANT,

YOU AUTOMATICALLY MAKE

THE 8-DOLLARS-AN-HOUR THAT PAYS

FOR THE BABY-SITTER

SO YOU CAN DANCE TABLE DANCES ON

AND UNDER THE TABLE, AND EVEN

HAVE A GREAT EXCUSE, ANYTIME,

NOT TO SHOW UP, TAX-FREE, TO WORK

BUT MOST STUPID OF ALL,

WE DON'T HAVE OR NEED JOBS

TO SUPPORT OUR CHILDREN.

THEY'RE FLOATING IN THE INTERNATIONAL

WATERS OF THE WORLD BANK

AND THE MONETARY FUND, RAISED BY EXPERTS,

THE SAME BREED WHO SAY WE NEED,

BUT WON'T FIND, ANOTHER SHAKESPEARE.

SHAMAL (94)

Of these places that are make-believe

SMALL WORLDS THAT WANT TO BE BIGGER

SHARING THE VICTORY (94)

I am sure 94

You are sure

CONKING HIM THROUGH HIS

SHATTERED WIG (94)

You claim the crack-dealers 94

As your army

I want mine hooked

On poetry

SHAW (94)

IN

SHAWNEE SILHOUETTE (92)

Death, Disease, Destruction: 91

One would think

Poets could get published

Before them

The black dog of my guilt 92

Eats your white cat

Nah, just kidding

AND

SHEILA NA-GIG (94)

Good poets need not say "fuck" 93

But what the fuck?

I always had problems with authority

If I do not wear a pink triangle 94

Does that mean I am not queer?

Am not hated?

Am not lied to?

Am not sent for and then sent away?

Am not asked to contribute my life?

HUM

"SHENANDOAH" (94)

Geraldine 94

Is a welcome

Re-leaf

AND

"SHIP OF FOOLS" (94)

If you had any idea of it, 94

You would not invoke

.... Poetry, so lightly

FOR A

SHOE TREE (91)

Toe pick 91

Tooth pick

Ice pick

Pick your pick

SHOFAR (94)

FAR FROM HOME

SHOOTING
STAR (94)

Sweating and spitting 91

Pages out of you

We, welfare mothers 94

Can not live on minimum wage

Turn the page, turn the page

SIDESHOW IN CRUEL
GARTERS (93)

Your poetry is soggy 93

With the sot-weed factor

MADE IN THE

SIERRA
NEVADA (94)

Did I hear you correctly? 94

Did you say

“Sing a song

And suck my cock?”

CLICHES, UNDERSTANDING THEM TO BE A

SIGNAL (93)

LIKE SMOKE

When the poem ends, 92

The world begins

Blood rains and tears rain 91

And coffee rains down drains

And love juice rains

And we dance in the rain

NEAR A

SIGNPOST (49th PARALLEL). (94)

Dead kisses 92

the

Head misses

Dread pisses

on

Shed promises

Milking poetry's shriveled breasts 91

The poem does not know

It's sucking on a wet-nurse

While its mother is full,

Ripe, exploding,

But too busy

Wildlife Loving 94

Scamper Camper

I fire the woods

To make sure

The animals

Are not dead

Ram stall 93

Butt out

HERE ARE THE

SILVER
APPLES (94)

WITH SILVER WINGS

Bitter scrabble 94

Letters that

Won't co-operate

Sky, sky, sky 91

GOSSAMER

SILVER WINGS (94)

What is time 94

Called in

Heaven?

Where is 92

God

Not manifest?

Are here and now 91

Only a dead-end intersection

To the hereafter?

Then kiss 93

His feet

EATEN BY

SILVERFISH (94)

Uncle memories 94

What did you make of them?

Occupational poetry? 92

Butcher and barber poetry?

What, of the above, are not metaphors

THAT

SING TO THE HEAVENLY
MUSE (94)

All boundaries are lies 91

And the bullet is a lie

And its whistling is a lie

And the pain is a lie

And falling down

Inside some boundary is a lie

The spirit of her tits 94

Took

A rest

YOU

SINGING
HORSE (94)

OF THE

SINGLE SCENE

Get screwed.

Just loving is always enough

To assuage any guilt at its loss

SINGLELIFE (91)

You gave your life to 91

The dream you dreamt of me

I did too

So sorry

IN

SINGULAR
SPEECH (94)

I'm going to do it again. 92

Hope you get fooled this time:

Irises

Moon like an opening jar, 94

Flowers (let's cut to the quick)

And death

Grave robbers! 91

I am a bit trite 93

As though my worms rain pockets of pages

As though the sun is rising in the feast of potatoes,

like butter

As though the flowers are brooming up bumble-trees

PATTERNS OF

SINISTER
WISDOM (93)

I whisper, 93

“I don’t care what you call yourself

If you’re not my own little twat-licker

And if you are, the victory

Is not in saying you’ve been there, done that”

ABOUT

SISTER VISION (94)

AND

No, it's not true; 94

Women of color do not have longer tongues

SISTERS
TODAY (94)

HUSH CORPORATE

A sound! 91

A cellophane sound,

But, yet, a sound.

Father's Lesson 94

All love is precious

SISYPHUS (93)

SISSY PUS

Bitter about the bitterness? 94

But you don't feed your man on the right diet

Of love and exercise and tenderness

Ah, well, who needs sugar? You can always buy it.

BUT WATCH OUT FOR

SKAZ (93)

SCAGS AND GERMS

SKOOB (94)

SCUBA DIVING FOR MUFFS

This paper is what I love least: 94

Stealing poems from my imagination

WITH

SKYLARK (94)

TONGUES

Damn Yankees! 93

Desperation Crocuses 92

Planting early

To remind us of life,

Hidden under Winter

(`Cus we can't seem to remember

There's life if we don't see flowers)

There is no "other side" to my biography. 94

I'm open to all

Dawn 91

Arrived

Like a smile

(But not like a simile)

SKYLARK (92)

WING BRUSHES

Put down pudding 92

Because you don't like

Sugar

THE

SKYLINE (91)

ON A

SLANT (92)

Sheltering the Anemone 92

Go on, eat me

When there's nothing left,

You'll move on

NOT WORRYING ABOUT

SLATE AND
STYLE (94)

If I could paint 94

I wouldn't waste all my fuckin' time with poetry

They Steal the Oil From the
Land 93

So I drive to the post office

Twenty times a week,

Mailing out letters of protest

In the fever of Summer 91

I found a pool of water

Jumped in with my friends

Shouted delight

Then Carl Rowan came out and shot me

THAT IS

SLIPPERY WHEN
WET (94)

IN THE

SLIPSTREAM
AUDIO (94)

All my poems suck 93

I was told I could get 92

My old teacher's job

If I reported him

For child abuse

In toilet tissue details 91

The big black boot

Of injustice is spared

A forest review

To That Lazy Librarian 94

If you'd been doing your job

She wouldn't need picture-books

About pregnancy

She would have read from lines like mine

About what a GRAND time

It is to be on your knees, pumping out sperm,

On your back, ingesting injustice

And how we're made to pay and pay and pay

For the pleasure

OF A

SLOUGH (91)

SLOW
DANCER (92)

At my poetry readings 91

I do a table dance to pronounce

The beauty of the line

Of my legs and hips

To accentuate the body of my work,

Because men (and most women) will listen to nipples

Stuck out in the middle of a tee-shirt line

SMALL POND (94)

Grandma wore a live fox as a scarf 92

SMALL PRESS (94)

And little lines of poetiferousness, too 91

SMALL TIMES (92)

Living in Reality and Dreams 92

Broken

Fantasy

Breaker,

Broke

THE

SMITH
GENERALIST (94)

Forever frozen in
mid-sentence 94

Kick-boxing karate angels

I will tell you that I really --.....

LOST THE

SMITH
PEREGRINE -- (94)

BARTH'S "POETIC PARAGRAPH"
EXTENDED INTO A BOOK, BY GOSH!

LEAVING

SMITH'S KNOLL (94)

Little boys 93

Skinny dipping

In your pool

In secret

Moonlight

Gotta prepare 94

For pederasts

Like you

BEHIND

SMOKE:
WINDOWS (92)

WITHOUT

SMOKE
SIGNALS (93)

I crawl into the horn of her
trumpet 93

To reach the stars

Gotta care -- 91

To crack

That earth

One more time

AT A

SNAIL'S PACE (94)

Consider the options, 94

What you could be doing, right now:

Curing cancer

Creating a world

Resting your brain

Licking my bosom

LIKE A SLOW

SNAKE (94)

Love 92

Does not always take on

A particular tone,

Gives in peculiar,

Unexpected ways

IN A

SNAKE RIVER (94)

MISSING THE

SNOWY
EGRET (94)

EFFECT

Let me blow my hot air 94

Over your

Conifers

Candles crack 91

Perfect darkness

Gasses shine

In the surround

OF THE COMPANION CATTLE OF

SOCIAL
ANARCHISM (94)

AND ANACHRONISM

What people do you
remember? 92

The ones who slurped and burped?

Or the weird ones that they wanted to be?

SOCIAL JUSTICE JOURNAL OF CRIME,
CONFLICT, WORLD ORDER (92)

And prison was all about brain
scrambling 92

Before they gave up and called a trash
strike

And threw rotten eggs and sticks and
bombs

IN THE

SOCIETY (94)

Battered 92

War and pancakes

Withered 94

Before mankind was so terrible

So hopeless and hollow

You heard the tones of “in whose image”

OF

SOJOURNERS (94
)

Earth 94

Earth

Earth

Poet slips in

A spirit

UNDER FRENCH TOAST

I don’t even see 92

Yesterday

Though these tears

SOLEIL (94)

EACH TRYING TO BE A

SOLO FLYER ON SPARE
CHANGE (94)

Mutant tornado 92

With double vision

The poem without words 94

Said

Everything I was going to say

DREAMS THAT

SOME BEES (93)

CALL FOR THE HONEY OR

THE
(SOMETHING) (91)

(Homunculus said that) 91

SONOMA
MANDALA (94)

MANDELA MANIA (GREAT OR NOT)

Necklaces 94

There are people who are combustible

We know this because our solid waste policy

Allows the burning of tires

EAST OF

SONORA (94)

Amazing -- 94

How the architect of your house

Assumes

That you are not butter

Sucking bronze statues, 92

I lose the flesh features

Of the man that he is,

Deny him his humanity

By denying the truth

Of beauty

Objectifying metal

Like men aren’t able to --

They’re the romantic ones.

I don't care about flowers,

Just like to watch them pose

And crawl

PLUMBING THE DEPTHS WITH

SOUNDINGS FOR SURVIVORS OF CHILDHOOD SEXUAL ABUSE ECHOES NETWORK (93)

No one was ever nice to
me -- 93

Not the way I wanted them to be

SOUNDINGS
EAST (94)

Muscatine omniscient
voice 92

And such limited knowledge of she, they,
you, it

Anyone, anything

SOUNDINGS ON THE EDGE CITY, LIVING
ARTS (93)

Inbreeding keeps kids from having mad
eyes? 93

SOUNDS OF
POETRY (91)

The magic snowflake of art 91

Melts when the drummer

Stops sweeping his jazz-brushes

Across the roof

MAN!

SOUTH ASH (94)

Poetry Calm 94

Moon-willow words lain out over the shadowy swan,

Perfect pitches of serenity

To avoid -- ANY

Fuckin asshole coming in and upsetting the fuckin balance

Allright?!

…….Okay,

Swan fluffs its wings into midnight silence

EAST OF

SOUTH
CAROLINA (94)

LINGUS

Living on the faultline 92

Of a fart

LANGUISHING ON THE

SOUTH
COAST (94)

Seeds sown in our slippers 93

We throw out old laundry

Check the pockets of our jeans

For loose grandmothers

Hava Nagila? 91

Go ahead,

Have one on me

Unsulky Sue 94

Licking all the sweetness

Off of

Your sugar-daddy's lust

Like The Last Defender 92

He waited in my fox hole

All day

Knew of nothing else to do

WHEN NOT IN

SOUTH
DAKOTA (94)

OR KISSING THE GROUND OF THAT

SOUTH END (94)

OF

SOUTH
FLORIDA (92)

Wow, man! 92

Look! See how skillful

The birds are at flying, man?

Wow!

OR GETTING

SOUTH HEAD (93)

Hey, duck, 93

Duck!

(I think I need the other stanzas,

The north side of something, to the left of east)

MIXED UP WITH THE

SOUTHEASTERN FRONT (94)

OF

SOUTHERN CALIFORNIA (94)

Boulevard, boulevard 94

All my lines are boulevards

Or have a boulevard

Roses, Rues, and
Ruses 92

Bouquet in arms in crinkly paper

We wind our way down French boulevards

With names like Sacre CWWWWWWWHIC!

………… Cwwwwwwwwwwwhic! ….

CWWWWWWWWHHHIC!!!!
(Sorry -- hairballs)

FOR

SOUTHERN
HUMANITIES (94)

AS OPPOSED TO NORTHERN HUMANITIES

SOUTHERN
POETRY (94)

A good mount 92

Does not hang, after being hunted down,

Dead, against the wall,

Sometimes, isn't even horny

SOUTHERN (93)

I ask men for their secrets 93

To expose them

Before they expose themselves

Tee-shirt Poetry 92

"Up mine!"

SOUTHERN (94)

When is poetry ever like this -- 94

As dull and as commonplace

As zaz?

SOMETHING OR OTHER

Oh me, oh my 94

Oh, why isn't life

As dreary and dull

And awful as it is in books

Of poetry?

You have to admit 93

Everyone is crazy,

Completely insane,

Hopelessly lost in the ozone,

So, let's fuck, huh?

SOUTHERN ROSE TAPESTRY NETHER WORLD SHAMROCK (91)

Ill, 91

I'll

Will

Will

To Willy

Who will

Love me for it.

Or will he?

WHICH IS NOT CLOSELY ENOUGH DEFINED FOR THE

SOUTHWEST (94)

Weighing the morning 92
On the subjective scale
Of yesterday

OR, IDIOMATICALLY, THE

SOU'WESTER (94)

If morning is the matrix 91
Dusk is the determinant

Picture this one; 94
I envied the bitch in heat:
To be a bitch so openly
And, yet, be surrounded

I'll describe a scene 93
To trigger visions
So that you'll listen,
Like the puppy licking my chocolate breasts,

To the last line, which I will

Throw in as a statement.

Live free or die.

SOUNDING LIKE A SOW INVESTOR WHO FALLS FOR A

SOW'S EAR (94)

Perfection 92

Alberto the Cuban rebel/guitarist

Taught me how to split a log for the fire

(9 millimeter bullets are expensive and fun but

also efficient)

She raised her son 94

To be pussy-whipped

Heavy

Cream curdling in its container

PECKED, LIKE A HEN, THROUGH

SPACE AND
TIME (92)

BY A

Beach in Soggy Sand at Low
Tide 92

When I want you to focus on that solitary
footprint

I have to compromise and stain it with mint

SPARE TIME (92)

SPARROW WITH POLITICALLY INCORRECT
VERSE POVERTY PAMPHLETS (94)

Animal Secrets 91

Swallow, Tern, Robin,

Bitch, Ass, Maggot

That man with blond hair 92

Bought a sport's car

With my money and ran,

Was riding in the rain

With his new girl, Jane,

Cruising along, under

A railroad crossing.

In the glare of night

He didn't see quite right

Didn't notice the underpass

Was overfilled

With three feet of water

I guess that taught him,

Maybe even taught her.

The car came bubbling

To its eternal rest.

She hit her head.

(I liked that best)

Now, you see?

That was poetry.

Thing

Hefty bags to cart away all the

Rejections

Carrot peelings that were

Poetry

Blue bathroom tissue that was used

To wipe

Puddles of tears

After learning about

Death

Just a little

Looking up at me, lovingly,

While licking my thighs

FROM THE ENTIRE

RANGE OF

SPECTACULAR
DISEASES (94)

THE ENTIRE

SPECTRUM (94)

OF THE

SPECTRUM IN OLD
TIMES (94)

OF

SPIDER EYES TABOO
TALES (91)

SPINDRIFT (94)

SPINSTER'S INK (92)

STAINS

Now sheets(ironed) are
“blank” 92

Not crisp

And old ladies that bore

Generations/ (Bore readers,) get

Only “an hour’s rest”?

(I picture them, lost at their typewriters

Spitting on their grandmother’s legacies

And my grandmother’s too.

You spit on your grandmother’s

Legacy!

That’s why you’re lost;

You don’t know the synthesis of rebellion

The smushing together of pursed lips

The iron is still too cool

To make a good sizzle

THE

SPIRIT THAT
MOVES (94)

"What have I done for me
lately?" 94

That's what a true woman of color would say

In my cataract world 92

All highlights

Are fresh-fallen snow

IN A

SPIRITUAL
QUEST (91)

Spiritual quest 91

For punctuation to

Untie unintended

Ambiguity

FOR

SPIT (94)

You read these lines, nothing. 94

When the horizon is bigger than the moon 92

Can you see the crack

In the atmosphere?

Can you see

The brown asshole of fluorocarbons?

TO MAKE A

SPITBALL (94)

Hey, honey, 92

Your eyes are deviled eggs

You ought to see a doctor

CROWD WHO ARE

CAUGHT BETWEEN THEIR

SPLIT
PERSONALITY (92)

Schizophrenic Poets and Religious Lovers 92

Hating progress

Loving the fruits of it

Sucking the juice out of their word-processors

SPOKES (94)

TURNING TABLES SET WITH A

SPOON RIVER (94)

No man 94

Has had his hands up my dress

And down into the elastic of my pantyhose

For years and years,

Since one, (that one, that time),

Who was folding my clean laundry for me

Asked, stretching those plastic containers of leg-air,

"Why do you wear these stupid things?"

No good answer came and until one does,

I won't.

Treesap gold 92

Outside darkness

Memory exercises

For potatoes without couches

SQUEAKY
WHEELS (92)

Squeak, squeak 92

Squeak,

Squeak

STAND (94)

Man, 92

This ain't right, man.

I don't want no fuckin'

Hang gliders

Scaring my fuckin' birds

BEFORE THE

STANDING
STONE (92)

STAPLE (94)

Staple this to your brain: 93

You can call the cat with poetry

We stole a cadaver's arm 94

From Yale,

Passed more than a quarter

To the poor man working the toll-booth,

That night

Scat John! 92

I've had men like you, John.

I usually stick your faces

In it

Then say, "Bad boy!"

And make you clean up your mess

Before I kick you out the door

STAR (94)

CHARTS

How can I tell the loss of a child 94

From the loss of a lover

When, imagining the child,

I see him as a man grown tall?

I guess maybe I wasn't supposed

To be a mother.

Hey, Norma, dear, 91

I see you there,

Sitting on your ample

Peripheries

ON

STAR ROUTE (92)

When I see the pear, 91

Passed over the counter,

I see…..

Ah, shit, I can't lie --

I see a pear passed over the counter

STARLIGHT (92)

TWINKLES

"There's a woman, Ira!" 92

I say. He says

"Ah, shit, why bother?"

I agree, leaves more women

For me. Let Ira have coitus

In the open womb

Of his opulent room,

In the cold pit, the dry

Wit of his poetry

IN

STARMIST (94)

Treasures? 92

Don’t natural wonders display their own price tags?

In Another Mist, and Not a Missed Mist 94

I’m sitting

Disturbed

Pressure against

My buttocks

And I’d be happy

To pinch a loaf

But don’t

I need an image, a new image 93

I need imagery, NOW,

Or I’ll go all prosey

Uh, uh

Petunia in a pancake

Uh, uh

Six galloping julien fries

Uh, uh

The moon, sandwiched between lettuce and bacon

Uh, uh

Ice cream dripping onto my sweating cleavage,

Strawberry bosom, scooped by tender tendrils of butterfly

Shrimp --

Oh! ….

Uh, nevermind.

Maybe all I needed was lunch

STARSONG (92)

Moron of many big words 91

He looks like a zoot-suiter 92

Spinning keys on a chain,

That precocious little baby of mine,

Playing jump rope with our umbilical cord

STATE
STREET (94)

Once there was no imagery worth
naming, 92

We had to resort to video

STEELHEAD
SPECIAL (94)

The game warden And
poacher 93

Entertain each other

Sea-Sick Fishing Chum 94

Fishing for poetry,

He vomits imagery,

Bait

STEREOPTICON (92
)

Crawfish crayon 92

Rolls over

The desk's edge

STEVAN (94)

STEVENS (94)

Chalk crumbles to bits 92

In the excited teacher's hands

We, the experienced, know the value of chalk

Occurred to me that poetry is not 94

In the tone of the voice, song it may be

But just the fact that lines are measured

Without conveyance of feeling, thought,

Philosophy even, as long as it's strong

In what it conveys and meticulous

About word choice. I think it must

Have a certain mystery about it,

How simple words can shine so

STICK
LONGHAND (92)

LINE

No embers, here 92

STICKS (94)

Puke, puke, puke 92

You don't have to puke

Up poetry

That's what all the guys say, 93

"Tastes good! Try it."

I say, "You try it."

Worship his penis 94

The ultimate lie

Both heads will explode

And then he'll die

STILL WATERS (94)

Hey, ice-hole 93

That's some bitchin' lichen

Ooooops! I must end up

Liking lichen

Hey, don't sing bird songs to a cat like
me 94

Men tell me that, too 92

"My fingers will set you on fire"

But I'm water. I flow through the cracks

Of your luke-warm deceit

STOHLMBERG (94
)

You 94

Really mean nothing to me

I'd rather have a revolver

Than your face-down, flat poetry

TO THE

Third world soup song

Everyone should eat soybeans

I don't need steak

Anyway, I'm on a starvation diet

OF PEELED

STONE (94)

INTO

STONE SOUP (94)

THEY HAVE US COMPETING

IN DEAD FIELDS

LIKE POETRY

(THOUGH TEACHING AND LECTURING MAKE MONEY)

THEY HAVE US COMPETING

COMPETING AGAINST THE BODY

THAT USED TO BE "OURSELVES"

THEY HAVE US COMPETING --

NOT THE GOOD KIND, EITHER,

FOR TOO FEW PRIZES

FOR ALL SISTERS TO SHARE IN

JUST ENOUGH TO KEEP US,

SISTERS AND BROTHERS

FROM COMPETING AGAINST THEM

AND WHO IS “THEM”

IF ANY ONE OF “US” KNEW THAT

WE WOULD BE OUTCAST FROM “OURSELVES”

AND WHAT IF ALL THIS IS PARANOIA?

IF THERE IS NO “THEY”, NO CONTROLLING

ENTITY THAT WE ARE NOT,

THAT WOULD BE THE BEGINNING OF PARANOIA

It is easier to pen silver lines 92

When you’re born with a spoon in your mouth

When the blackberry moon
rises 94

When the grapefruit moon rises

When the aspirin moon rises

When the crescent moon rises

When the dixie-cup moon rises

When the donut moon rises

When the pastry moon rises

The pepsi moon

The blood-shot moon

The bean-dip moon

The umlaut moon

The incandescent moon

The silk-stockinged moon

The sugar-cane moon

When the mayonnaise moon rises

When the star-kissed moon

The bumble-bee moon

The tunafish moon

The albacore moon

The dolphin moon

The blowfish moon

The whaler's moon

When the worn-out moon

When the pizza moon

When the pock-marked moon

The chocolate moon

The cold-coffee moon

The constant moon

When the dental floss moon

When the breakdown moon

The morning moon

When the toilet paper moon rises

When the moon rises

I'll be there

Waiting for your reflection

Still pulled, like the water

By a tide of memory.

FOR

STONEVALE (94)

I fell asleep in the car 94

Woke up with two men

mauling me

They were NOT embarrassed

STOP LIGHT (92)

ENERGY

She's a poor girl, but rich
inspiration 92

For my, sometimes, slumming poetry

ALONG THE

STORMLINE (94)

OF

STORY POEMS (91)

STRAIGHT AHEAD (91)

"Hello, my friend, so good to see you again 91

I been down so long, all by myself

I just couldn't make it. Keep on straight ahead"

DEAD DIAGONALS OF

STRAIGHT
STANDARD -- (94)

Radical Lonna 94

19-year-old poet

Who says she is lonely

Protests nothing

We ought to take a trip together, 92

Maybe to the beach.

THE STRAIN (94)

IS THAT

STREET NEWS (94)

“Kill My Landlord” 94

`Cus Cambi says we got mushroom clowns

Growing up our asses

(or was it blowing up our asses?)

And tyrant whales about to overthrow us

And just enough air to breathe

To pay off this month’s thrills

FROM THE

STREET (94)

STRIDE TAXUS APPARITIONS OF
TROMBONE (94)

SLIDE THE CONSONANTS

INTO A POND OF FLOATING CODAS

FOR THE

STRUGGLE, THE PROLETARIAN REVOLUTIONARY (94)

Does socialism mean that all those
young 91

Spikes get an equal share of my pussy?

To each according to need?

From each according to ability?

Would my pussy be under state control?

"Oh, it's like that everywhere," I'm told.

Peace 92

Is the Pong

Of the diplomatic

We arrest you for having "no I.D." 93

That sounds like fascism to me.

I defend the public 94

With my degree, my wits

And 27 years of T.V.

ASPIRING TO

STUDENT
LEADERSHIP (94)

IT IS ALL ABOUT POWER AND ACCESS --

Peace but no fried ship 91

Shards ripped the thing unborn

A can of balls? 92

Are they orange or yellow?

Are they still fuzzy?

Can I play with them?

There's vision's glare 94

And there're sunglasses

Charting shards 91

In proportional integrals

ARE YOU REALLY AN ARTIST WHEN YOU
GET A

STUDIO (94)

OR

Copper grows from our mother's
ass 92

TWO STUPID PLACES

WHERE YOU CAN PLOW YOUR MIND'S

SUB-TERRAIN AND
ANVIL (94)

SWORDS

I really don't need to read the poem -- 94

Fuck, I saw the movie

FROM

SUCARNOCHEE (94)

I write these lines to tell 94

You I had a bad year,

Feel like a paint-by-number

Cardboard clown

With the colors mixed up

INTO

SULPHUR (94)

Brimstone dream invasions -- 94

Is it the city or some asshole?

Nan's Last Sneeze 92

Blew her brains out

Busted a blood vessel

The ears go first....

....eh?

EVEN TEMPERED IN A

SULPHUR RIVER (94)

He farts in perfect pitch 94

Arthritic fingers tap the table

Music comes from the strangest depths

Prune juice adjusts timbre

Chair cushion controls amplitude and duration

In old age, he has developed

A new music of all the senses

THAT GREW FROM A

SUMMER STREAM (94)

Made me simile 92

Made me smile

Made me like simile

Made me like simile

Made me smile like simile

Made me like smiling like simile

Made simile like a smile

Empathy 91

Burnout and not a peep from the

Pope

OVERHUNG BY

SUN (94)

COULD NOT MAKE ART OF THE

SUN DOG
SOUTHEAST (94)

WHO LIES IN THE NORTHERN

SUNSHINE READING GOOD
READING (93)

God's gone 93

On vacation

During dreary days?

Bud trees the oh! 92

ABOUT THE

SUPERINTENDENT'S PROFILE AND POCKET EQUIPMENT DIRECTORY. (94)

SWALLOW! --WHAT A MAGAZINE NAME! -- AT LEAST IT'S

TRUTHFUL ABOUT WHAT MANY MEN MEAN "SWALLOW, BITCH"

WHAT A SWALLOWING

SWAMP ROOT (93)

Kleenex bubbled a bit green 92

Three large specks, grey-brown

Two little ones

Four specs had nose-hairs sticking out

SWORD OF
SHAHRAZAD (92)

I’m menstruating -- 92

You want to have sex?

HE REALLY IS PRETENDING TO BE, A

SYCAMORE (94)

I grew a dang pickeral-style 91

pickle this big!

Nothing happened 92

Didn’t cut my finger

There’s no way around it,

Three bottles

left

Little mushroom 94

Cloud

Where minnow

Used to be

SYZYGY IN THE IMPOSSIBLE UTTERANCE OF A PLASTER CRAMP(92)

Made in the USA
Monee, IL
03 November 2026

67385828R10192